VERIFYING GREENHOUSE GAS EMISSIONS

METHODS TO SUPPORT INTERNATIONAL CLIMATE AGREEMENTS

Committee on Methods for Estimating Greenhouse Gas Emissions

Board on Atmospheric Sciences and Climate

Division on Earth and Life Studies

NATIONAL RESEARCH COUNCIL
OF THE NATIONAL ACADEMIES

THE NATIONAL ACADEMIES PRESS
Washington, D.C.
www.nap.edu

THE NATIONAL ACADEMIES PRESS • **500 Fifth Street, N.W.** • **Washington, DC 20001**

NOTICE: The project that is the subject of this report was approved by the Governing Board of the National Research Council, whose members are drawn from the councils of the National Academy of Sciences, the National Academy of Engineering, and the Institute of Medicine. The members of the committee responsible for the report were chosen for their special competences and with regard for appropriate balance.

This study was supported by the United States intelligence community. Any opinions, findings, and conclusions, or recommendations expressed in this material are those of the authors and do not necessarily reflect the views of the intelligence community.

International Standard Book Number-13: 978-0-309-15211-2
International Standard Book Number-10: 0-309-15211-9
Library of Congress Control Number: 2010926783

Copies of this report are available from the program office:
Board on Atmospheric Sciences and Climate
500 Fifth Street, N.W.
Washington, DC 20001
(202) 334-3512

Additional copies of this report are available from the National Academies Press, 500 Fifth Street, N.W., Lockbox 285, Washington, DC 20055; (800) 624-6242 or (202) 334-3313 (in the Washington metropolitan area); Internet, http://www.nap.edu

Cover: Major sources and sinks of anthropogenic greenhouse gases. Cover design by Van Nguyen.

Copyright 2010 by the National Academy of Sciences. All rights reserved.

Printed in the United States of America

THE NATIONAL ACADEMIES
Advisers to the Nation on Science, Engineering, and Medicine

The **National Academy of Sciences** is a private, nonprofit, self-perpetuating society of distinguished scholars engaged in scientific and engineering research, dedicated to the furtherance of science and technology and to their use for the general welfare. Upon the authority of the charter granted to it by the Congress in 1863, the Academy has a mandate that requires it to advise the federal government on scientific and technical matters. Dr. Ralph J. Cicerone is president of the National Academy of Sciences.

The **National Academy of Engineering** was established in 1964, under the charter of the National Academy of Sciences, as a parallel organization of outstanding engineers. It is autonomous in its administration and in the selection of its members, sharing with the National Academy of Sciences the responsibility for advising the federal government. The National Academy of Engineering also sponsors engineering programs aimed at meeting national needs, encourages education and research, and recognizes the superior achievements of engineers. Dr. Charles M. Vest is president of the National Academy of Engineering.

The **Institute of Medicine** was established in 1970 by the National Academy of Sciences to secure the services of eminent members of appropriate professions in the examination of policy matters pertaining to the health of the public. The Institute acts under the responsibility given to the National Academy of Sciences by its congressional charter to be an adviser to the federal government and, upon its own initiative, to identify issues of medical care, research, and education. Dr. Harvey V. Fineberg is president of the Institute of Medicine.

The **National Research Council** was organized by the National Academy of Sciences in 1916 to associate the broad community of science and technology with the Academy's purposes of furthering knowledge and advising the federal government. Functioning in accordance with general policies determined by the Academy, the Council has become the principal operating agency of both the National Academy of Sciences and the National Academy of Engineering in providing services to the government, the public, and the scientific and engineering communities. The Council is administered jointly by both Academies and the Institute of Medicine. Dr. Ralph J. Cicerone and Dr. Charles M. Vest are chair and vice chair, respectively, of the National Research Council.

www.national-academies.org

COMMITTEE ON METHODS FOR ESTIMATING GREENHOUSE GAS EMISSIONS

STEPHEN W. PACALA *(Chair)*, Princeton University, Princeton, New Jersey
CLARE BREIDENICH, Independent Consultant, Seattle, Washington
PETER G. BREWER, Monterey Bay Aquarium Research Institute, Moss Landing, California
INEZ FUNG, University of California, Berkeley
MICHAEL R. GUNSON, Jet Propulsion Laboratory, Pasadena, California
GEMMA HEDDLE, Chevron Corporation, San Ramon, California
BEVERLY LAW, Oregon State University, Corvallis
GREGG MARLAND, Oak Ridge National Laboratory, Oak Ridge, Tennessee
KEITH PAUSTIAN, Colorado State University, Fort Collins
MICHAEL PRATHER, University of California, Irvine
JAMES T. RANDERSON, University of California, Irvine
PIETER TANS, National Oceanic and Atmospheric Administration, Boulder, Colorado
STEVEN C. WOFSY, Harvard University, Cambridge, Massachusetts

National Research Council Staff

ANNE M. LINN, Study Director, Board on Earth Sciences and Resources
JANEISE STURDIVANT, Program Assistant, Board on Atmospheric Sciences and Climate

BOARD ON ATMOSPHERIC SCIENCES AND CLIMATE

ANTONIO J. BUSALACCHI, JR. (*Chair*), University of Maryland, College Park
ROSINA M. BIERBAUM, University of Michigan, Ann Arbor
RICHARD CARBONE, National Center for Atmospheric Research, Boulder, Colorado
WALTER F. DABBERDT, Vaisala, Inc., Boulder, Colorado
KIRSTIN DOW, University of South Carolina, Columbia
GREG S. FORBES, The Weather Channel, Inc., Atlanta, Georgia
ISAAC HELD, National Oceanic and Atmospheric Administration, Princeton, New Jersey
ARTHUR LEE, Chevron Corporation, San Ramon, California
RAYMOND T. PIERREHUMBERT, University of Chicago
KIMBERLY PRATHER, Scripps Institution of Oceanography, La Jolla, California
KIRK R. SMITH, University of California, Berkeley
JOHN T. SNOW, University of Oklahoma, Norman
THOMAS H. VONDER HAAR, Colorado State University/CIRA, Fort Collins
XUBIN ZENG, University of Arizona, Tucson

Ex Officio Member

GERALD A. MEEHL, National Center for Atmospheric Research, Boulder, Colorado

National Research Council Staff

CHRIS ELFRING, Director
LAURIE GELLER, Senior Program Officer
IAN KRAUCUNAS, Senior Program Officer
MARTHA McCONNELL, Program Officer
MAGGIE WALSER, Associate Program Officer
TOBY WARDEN, Associate Program Officer
JOSEPH CASOLA, Postdoctoral Fellow
RITA GASKINS, Administrative Coordinator
KATIE WELLER, Research Associate
LAUREN BROWN, Research Associate
ROB GREENWAY, Program Associate
SHELLY-ANN FREELAND, Senior Program Assistant
AMANDA PURCELL, Senior Program Assistant
RICARDO PAYNE, Program Assistant
SHUBHA BANSKOTA, Financial Associate

Foreword

Agreements to limit emissions of greenhouse gases are a current focus of international negotiations. Such agreements are sought partly because gases emitted from each nation spread globally in the atmosphere—a relatively inert gas spreads quickly in an east-west direction, then vertically and in a north-south direction. For gases with survival times of more than few years (such as carbon dioxide, nitrous oxide, methane, many fluorinated hydrocarbons, but not ozone), constant emissions in the northern hemisphere result in nearly equal atmospheric amounts both north and south of the equator after several years.

National targets for emissions are being discussed worldwide, as are baseline years against which changes are to be compared. But how well can we determine whether a nation is meeting its targets and how well do we know nation-by-nation emissions in baseline years, past or future? From my own research in atmospheric chemistry, I know that little research has been done to answer these questions. Similarly, governments have not focused much attention on how well these quantities can be estimated and monitored.

Physical scientists might assume that monitoring and verification would be based on physical measurements of greenhouse gases in air and water and processes involving soils and vegetation. Business leaders and diplomats might assume that self-reported data based on activities like fossil-fuel usage and other measures would be used for monitoring and verification. This report shows how data from both realms can be used and also how to improve the respective estimates.

The details of future agreements are not yet known; for example, will responsibilities apply to single nations or to multination regions? Will individual baseline comparison years be employed or will rates of decreases in emissions be specified differently? The authoring committee provides answers to some anticipated monitoring and verification requirements while also creating a framework from which specific information can be drawn, along with ways to improve scientific capabilities. Future agreements or carbon markets might include credits for uptake of carbon dioxide. This report's discussion of agriculture, forestry, and other land-use activities can lead to improved, scientifically based estimates of the accuracy of uptake rates and methods of monitoring them.

This study was initiated by the National Research Council because of our perception that the questions above had not received enough attention from scientists, engineers, or governments. The authoring committee could draw from a very limited literature on these subjects and some of their findings and conclusions are original.

The extent to which monitoring and verification requirements will be incorporated into future international agreements on greenhouse gases and/or carbon markets will be decided by political and business leaders. This report informs those communities and scientists as to our current capabilities and also how to improve those capabilities over time. The committee chair, Professor Stephen Pacala, the other committee members, Dr. Anne Linn, and the reviewers deserve our thanks for this excellent report.

Ralph J. Cicerone
Chairman, National Research Council and
President, National Academy of Sciences

Preface

Greenhouse gas emissions are estimated for a variety of purposes, including gauging the success of mitigation measures, conducting basic research on biogeochemical cycles, and carrying out agency operations. Such estimates are made for a variety of gases, at a variety of scales and with a wide range of uncertainties. With negotiations for a climate treaty under way, it is timely to ask how well the greenhouse gas emissions of individual countries can be monitored and verified, and what improvements can be made to support a treaty. The National Research Council's Committee on Methods for Estimating Greenhouse Gas Emissions was established to carry out the following study:

> The study will review current methods and propose improved methods for estimating and verifying greenhouse gas emissions at different spatial (e.g., national, regional, global) and temporal (e.g., annual, decadal) scales. The greenhouse gases to be considered are carbon dioxide (CO_2), chlorofluorocarbons (CFCs), hydrofluorocarbons (HFCs), nitrous oxide (N_2O), methane (CH_4), and perfluorinated hydrocarbons (PFCs). Emissions of soot and sulfur compounds along with precursors of tropospheric ozone may also be considered. The results would be useful for a variety of applications, including carbon trading, setting emissions reduction targets, and monitoring and verifying international treaties on climate change.

The committee met four times to gather input, deliberate, and write its report. After its last meeting, the committee prepared a letter report on the capabilities of CO_2-sensing satellites to monitor and verify greenhouse gas emissions. The most promising of these satellites—the National Aeronautics and Space Administration's Orbiting Carbon Observatory—had failed at launch, and a decision on whether to replace it was expected before the committee's final report was completed. A final decision on a replacement mission has not yet been made, and the information and analysis in the letter report are included in this final report.

The committee thanks the individuals who briefed the committee or provided other input: Fred Ambrose, Richard Birdsey, Mausami Desai, Leon Fuerth, Jeffery Goebel, Samuel Goward, Kevin Gurney, Bill Irving, Maurice LeFranc, Michael Levi, Hank Margolis, Paul McArdle, Gilbert Metcalf, Joseph Norbeck, Lee Schipper, Dale Simbeck, Karen Treanton, Riccardo Valentini, Rod Venterea, Wenwen Wang, Zhonghua Yang, and Linda Zall. The committee also wishes to thank the NRC staff in general and Anne Linn in particular for exceptional efficiency, supernatural patience, expert editing, and good humor. Her knowledge and assistance were critical and she made the process a pleasure.

Stephen W. Pacala, *Chair*

Acknowledgments

This report has been reviewed in draft form by individuals chosen for their diverse perspectives and technical expertise, in accordance with procedures approved by the National Research Council's (NRC's) Report Review Committee. The purpose of this independent review is to provide candid and critical comments that will assist the institution in making its published report as sound as possible and to ensure that the report meets institutional standards for objectivity, evidence, and responsiveness to the study charge. The review comments and draft manuscript remain confidential to protect the integrity of the deliberative process. We wish to thank the following individuals for their review of this report:

Scott Doney, Woods Hole Oceanographic Institution, Massachusetts
Emanuel Gloor, University of Leeds, United Kingdom
Richard Goody, Harvard University (emeritus), Falmouth, Massachusetts
Isaac Held, National Oceanic and Atmospheric Administration, Princeton, New Jersey
Richard Houghton, Woods Hole Research Center, Massachusetts
Ralph Keeling, Scripps Institution of Oceanography, La Jolla, California
Denise Mauzerall, Princeton University, New Jersey
Arvin Mosier, U.S. Department of Agriculture (retired), Mount Pleasant, South Carolina
Michael Obersteiner, International Institute for Applied Systems Analysis, Laxenburg, Austria
Paul Wennberg, California Institute of Technology, Pasadena

Although the reviewers listed above have provided constructive comments and suggestions, they were not asked to endorse the report's conclusions or recommendations, nor did they see the final draft of the report before its release. The review of this report was overseen by William L. Chameides, Duke University, and Charles E. Kolb, Aerodyne Research, Inc. Appointed by the NRC, they were responsible for making certain that an independent examination of this report was carried out in accordance with institutional procedures and that all review comments were carefully considered. Responsibility for the final content of this report rests entirely with the authoring panel and the institution.

Contents

SUMMARY	1
1 INTRODUCTION	11
Domain of the Report, 11	
Overview of Greenhouse Gas Emissions, 15	
Organization of the Report, 20	
2 NATIONAL INVENTORIES OF GREENHOUSE GAS EMISSIONS	21
Developing and Reporting National Inventories, 21	
Sector-Based Reporting, 24	
Limitations of National Inventories for Monitoring, 28	
Near-Term Capabilities for Improving National Greenhouse Gas Inventories, 33	
Recommendations, 35	
3 MEASURING FLUXES FROM LAND-USE SOURCES AND SINKS	37
Remote Sensing, 37	
Improving UNFCCC Inventories of Land-Use Emissions, 46	
Future (>5 Years) Opportunities and Threats, 49	
Recommendations, 50	
4 EMISSIONS ESTIMATED FROM ATMOSPHERIC AND OCEANIC MEASUREMENTS	53
Inverse Modeling Studies of Greenhouse Gas Emissions, 54	
New Approaches for Increasing the Accuracy of National Emissions Estimates, 59	
Recommendations, 68	
REFERENCES	71
APPENDIXES	
A UNFCCC Inventories of Industrial Processes and Waste	85
B Estimates of Signals Created in the Atmosphere by Emissions	89

C	Current Sources of Atmospheric and Oceanic Greenhouse Gas Data	93
D	Technologies for Measuring Emissions by Large Local Sources	103
E	Biographical Sketches of Committee Members	105
F	Acronyms and Abbreviations	109

Summary

The world's nations are moving toward agreements that will bind us together in an effort to limit future greenhouse gas emissions. With such agreements will come the need for all nations to make accurate estimates of greenhouse gas emissions and to monitor their changes over time. In this context, the National Research Council convened a committee of experts to assess current capabilities for estimating and verifying greenhouse gas emissions and to identify ways to improve these capabilities.

This report is focused on the greenhouse gases that result from human activities, have long lifetimes in the atmosphere and thus will change global climate for decades to millennia or more, and are currently included in international agreements. These include carbon dioxide (CO_2), methane (CH_4), nitrous oxide (N_2O), hydrofluorocarbons (HFCs), perfluorinated hydrocarbons (PFCs), and sulfur hexafluoride (SF_6)—all of which are covered by the United Nations Framework Convention on Climate Change (UNFCCC)—and chlorofluorocarbons (CFCs), which are covered by the Montreal Protocol. The report devotes considerably more space to CO_2 than to the other gases because CO_2 is the largest single contributor to global climate change and is thus the focus of many mitigation efforts. Only data in the public domain (available to all without restriction or high cost) were considered because public access and transparency are necessary to build trust in a climate treaty.

The report concludes that each country could estimate fossil-fuel CO_2 emissions accurately enough to support monitoring of a climate treaty (see Table S.1).

However, current methods are not sufficiently accurate to check these self-reported estimates against independent data (e.g., remote sensing, atmospheric measurements) or to estimate other greenhouse gas emissions. Strategic investments would, within 5 years, improve reporting of emissions by countries and yield a useful capability for independent verification of greenhouse gas emissions reported by countries. Table S.1 shows that by using improved methods, fossil-fuel CO_2 emissions could be estimated by each country and checked using independent information with less than 10 percent uncertainty. The same is true for satellite-based estimates of deforestation, which is the largest source of CO_2 emissions next to fossil-fuel use, and for afforestation, which is an important sink for CO_2. However, self-reported estimates of N_2O, CH_4, CFC, HFC, PFC, and SF_6 emissions will continue to be relatively uncertain and we will have only a limited ability to check them with independent information.

METHODS FOR ESTIMATING GREENHOUSE GAS EMISSIONS

The report examines three categories of methods for estimating greenhouse gas emissions: national inventories, atmospheric and oceanic measurements and models, and land-use measurements and models. Under the UNFCCC, countries are required to inventory the human activities that cause greenhouse gas emissions, such as fossil-fuel consumption or forestry, and then multiply each activity level by its rate of emissions (emission factor). Uncertainties in the self-

TABLE S.1 Current and Near-Term Capabilities for Estimating National Anthropogenic Greenhouse Gas Emissions

Gas	Major Sectors or Activities	Method	Current Uncertainty for Annual Emissions[a]	Possible Improvements in 3-5 Years	Uncertainty of Improved Methods
CO_2	Total anthropogenic	UNFCCC inventory	1 (developed countries)[b]	Adopt most accurate methods in all countries	1
CO_2	Fossil-fuel combustion	UNFCCC inventory	1-2 (developed countries)	Adopt most accurate methods in all countries	1
CO_2	Fossil-fuel combustion	Atmospheric measurements and models	4-5	Develop improved tracer-transport inversion through new observations ($^{14}CO_2$, additional ground stations, Orbiting Carbon Observatory [OCO] replacement) and data assimilation	1-3 (annual) 1-2 (decadal change)
CO_2	Large local sources (e.g., cities, power plants)	Atmospheric measurements and models	5	Develop and deploy a CO_2 satellite program, including an OCO replacement, new in situ measurements in cities, and a research program to guide network design and satellite validation	2 (annual) 1 (decadal change)
CO_2	Agriculture, forestry, and other land-use (AFOLU) net emissions	UNFCCC inventory	1-4 (developed countries)	Adopt most accurate methods and activity data; improved emission factors through research and comprehensive ecosystem inventories	1-3
CO_2	AFOLU	Atmospheric measurements and models	5	Develop improved tracer-transport inversion through new satellite and in situ observations	4-5
CO_2	AFOLU	Land-use measurements and models	2-4	Develop improved observations, data assimilation, and models with ecosystem research	2-3
CO_2	Deforestation and degradation source, afforestation sink	Land-use measurements and models	2-4 (forest area change) 3-4 (emissions)	Develop improved observations, Landsat continuity, data assimilation, and models with ecosystem research	1-2 (forest area change) 2 (emissions)
CH_4	Total anthropogenic	UNFCCC inventory	2-3 (developed countries)	Adopt most accurate methods and activity data and improved emission factors through research	1-3
CH_4	Total anthropogenic	Atmospheric measurements and models	3-5	Develop improved tracer-transport models, new satellite and in situ observations, and improved emission models through research	2-3
CH_4	Energy, industrial processes, and waste	UNFCCC inventory	1-5 (developed countries)	Adopt most accurate methods and activity data and improved emission factors through research	1-2
CH_4	AFOLU	UNFCCC inventory	2-4 (developed countries)	Adopt most accurate methods and activity data and improved emission factors through research	2-3
N_2O	Total anthropogenic	UNFCCC inventory	2-5 (developed countries)	Adopt most accurate methods and activity data and improved emission factors through research	2-4
N_2O	Total anthropogenic	Atmospheric measurements and models	4-5	Develop improved tracer-transport and emission models, additional observations	3-5
N_2O	Energy and industrial processes	UNFCCC inventory	3-5 (developed countries)	Adopt most accurate methods and activity data and improved emission factors through research	3-4
N_2O	AFOLU	UNFCCC inventory	2-5 (developed countries)	Adopt most accurate methods and activity data and improved emission factors through research	2-4
CFCs, PFCs, HFCs, and SF_6	Industrial processes	UNFCCC inventory	1-4 (developed countries)	Adopt most accurate methods in all countries	1-3

TABLE S.1 Continued

Gas	Major Sectors or Activities	Method	Current Uncertainty for Annual Emissions[a]	Possible Improvements in 3-5 Years	Uncertainty of Improved Methods
CFCs, PFCs, HFCs, and SF_6	Industrial processes	Atmospheric measurements and models	4-5	Develop gridded inventories, improved tracer-transport inversion, and measurement of correlated variations of gases	2-5

NOTES: 1 = <10% uncertainty; 2 = 10-25%; 3 = 25-50%; 4 = 50-100%; 5 = >100% (i.e., cannot be certain if it is a source or sink). Ranges represent emission uncertainties in different countries (e.g., 1-3 means that uncertainties are <10% in some countries, 10-25% in some, and 25-50% in others). Uncertainty levels correspond to two standard deviations. Shaded rows are the self-reported values; unshaded rows are the independent checks on the self-reported values from independent methods.

[a] Uncertainties for the magnitudes of decadal changes in national emissions can be computed from the numbers in the table using standard statistical methods. Decadal changes (the cumulative change in emissions over 10 years) are reported in the rows requiring OCO measurements because early estimation biases will be reduced in calculation of a decadal change. The uncertainty of a trend is reported as a percentage of the emissions at the beginning of the decade.

[b] Based on 2006 data reported by five developed countries (Denmark, Greece, Portugal, the United States, and Poland) with a range of institutional capabilities for compiling inventories. In countries where AFOLU sources dominate energy and industrial sources, the uncertainties for total anthropogenic emissions would be much higher.

reported national inventories depend on the data and methods used to create them, which in turn depend on each country's institutional and technical capabilities. In many developed countries, uncertainties are reported to be less than 5 percent for national CO_2 emissions from fossil-fuel use (Table S.1), which is the dominant source. With the exception of a few minor sources in the industrial sector, uncertainties are much higher for other greenhouse gases and sources and vary greatly by country. Uncertainties for the net CO_2 emissions from agriculture, forestry, and other land uses and for emissions of CH_4, N_2O, PFCs, HFCs, CFCs, and SF_6 from all sectors can be less than 25 percent in some countries and greater than 100 percent in others.

The second method for estimating greenhouse gas emissions, called tracer-transport inversion, is based on atmospheric and/or oceanic measurements of the gases and mathematical models of air and water flow. Tracer-transport inversion estimates the net sum of anthropogenic and natural sources and sinks. Uncertainties inferred from tracer-transport inversions are less than 10 percent for the net global CO_2 flux to the atmosphere but greater than 100 percent for anthropogenic CO_2 fluxes at national scales (Table S.1). These large uncertainties arise because of the small size of the anthropogenic signal relative to the large and uncertain natural cycles of emissions and uptake, errors in the reconstruction of atmospheric transport, and the paucity and limited distribution of observations. Tracer-transport estimates of emissions of N_2O, CH_4, and the synthetic fluorinated gases are currently too uncertain to verify national emissions.

The third method estimates emissions of CO_2, CH_4, and N_2O using methods that are conceptually similar to those used for UNFCCC inventories, but can be made using independent information on land cover. It can be used to estimate emissions from both natural sources and land-use activities, such as agriculture and forestry. Satellite imagery provides the remote information on land surface characteristics and change. This information is converted into estimates of emissions using biogeochemical models constrained by measurements of greenhouse gas exchange between the land and the atmosphere. Satellite imagery is particularly useful for constraining forestry activities and can be used to determine the area of deforestation and forest degradation. The total annual change in forest area has an uncertainty of 10-25 percent in northern forests and up to 100 percent in tropical forests (Table S.1). Uncertainties in emissions from deforestation, reforestation, and forest degradation are high for both annual values and trends, ranging from 25 to 100 percent, because of uncertainties in parameters used to translate deforested area into CO_2 emissions. Land remote sensing can also be used to estimate agricultural emissions by identifying the areas using certain agricultural practices, such as paddy rice. Annual uncertainties in CH_4 emissions from rice production are 25-50 percent, and uncertainties in N_2O emissions from synthetic fertilizer use and manure production are 50-100 percent.

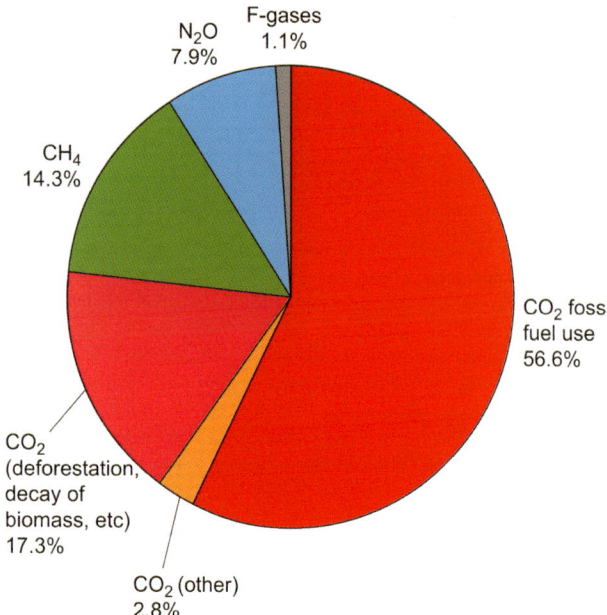

FIGURE S.1 Global anthropogenic greenhouse gas emissions and activities covered by the UNFCCC for 2004. These gases include CO_2, CH_4, and N_2O, as well as HFCs, PFCs, and SF_6 (the F-gases). The emissions of each gas are weighted by its 100-year global warming potential. Note that including short-lived greenhouse agents (e.g., ozone precursors) or decreasing the time horizon over which the global warming potential is calculated will decrease the fractional importance of fossil-fuel CO_2. SOURCE: Figure 1.1b from IPCC (2007b), Cambridge University Press.

Although uncertainties in emissions estimates are high overall, it may not be necessary to obtain accurate measurements of all greenhouse gases to support treaty monitoring and verification. The majority of anthropogenic greenhouse gas emissions covered by the UNFCCC are in the form of CO_2, primarily from fossil-fuel use (~74 percent in 2004; Figure S.1) and secondarily from deforestation (estimates range from 12 to 22 percent), making these two activities an obvious focus for monitoring.

FRAMEWORK FOR ESTIMATING AND VERIFYING GREENHOUSE GAS EMISSIONS

The UNFCCC framework for reporting national emissions comprises three main elements:

1. An internationally negotiated and accepted capability to monitor national, anthropogenic emissions of the most important greenhouse gases,

2. Independent review by an international body to determine whether appropriate procedures and methods are being used to prepare national inventories, to identify inconsistencies within and between reports, and to take action if problems are uncovered, and

3. An established mechanism through the Intergovernmental Panel on Climate Change (IPCC) to incorporate new information and strengthen inventory methods.

Because fossil-fuel CO_2 emissions can be estimated with reasonable accuracy using UNFCCC inventory methods (Table S.1) and because of the broad international support for the reporting framework, UNFCCC procedures have been, and will likely continue to be, the primary means for monitoring and verifying greenhouse gas emissions and reductions under a new international climate treaty. However, the current system has shortcomings for this purpose:

- Developing countries do not provide regular, detailed emissions reports.
- The availability of independent data against which to check self-reported emissions is limited.
- Estimates of CO_2 emissions from land use, as well as emissions of other greenhouse gases, have uncertainties that are greater than the expected emissions reductions over a treaty's lifetime.

The committee's recommendations are aimed at overcoming these weaknesses and improving the capability to estimate and verify greenhouse gas emissions in support of a climate treaty. Although some will take many years to implement, most will yield results within a few years.

RECOMMENDATIONS

The committee's recommendations fall into three broad categories: (1) strengthening national greenhouse gas inventories, which will likely remain the core of a global monitoring and verification system; (2) improving the ability to independently and remotely estimate national, annual fossil-fuel CO_2 emissions and to monitor emission trends; and (3) developing the capability to make accurate estimates of national CO_2, N_2O, and CH_4 emissions and CO_2 removals from sinks from agriculture, forestry, and other land uses, and to independently check self-reported estimates of CO_2 emissions from deforestation, reforestation, and forest degradation.

Strengthened National Greenhouse Gas Inventories

The two recommendations below are intended to improve the accuracy of national inventories and facilitate comparison with independent methods. The first recommendation would extend reporting of UNFCCC inventories to every country, and the second would increase the spatial and temporal resolution of emission estimates.

Recommendation. UNFCCC parties should strengthen self-reported national emissions inventories by working toward

- **Extending regular, rigorous reporting and review to developing countries, and**
- **Extending top-tier (most stringent) IPCC methods to the most important greenhouse gas sources in each country.**

UNFCCC reporting guidelines differ for developed (Annex I) and developing (non-Annex I) countries. Annex I countries report annual estimates for all anthropogenic sources and sinks of six greenhouse gases (CO_2, CH_4, N_2O, SF_6, PFCs, and HFCs) and a time series of annual estimates going back to 1990. In contrast, non-Annex I countries are required to produce only a periodic inventory of CO_2, CH_4, and N_2O at the sector (i.e., energy; industrial processes and product use; agriculture, forestry, and other land use; waste) level, without a detailed source breakdown. Most developing countries have produced only one national inventory to date. Financial and technical assistance will be required for developing countries to build an ongoing capacity to collect, analyze, and report emissions information regularly. Significant improvements in national inventories from 10 of the largest emitting developing countries (e.g., China, India) could be achieved at relatively modest cost (about $11 million over 5 years).

Countries choose among three tiers of methods for calculating emissions and removals of greenhouse gases. The lowest-tier methods (Tier 1) are simple and use default values for emission factors. Tier 2 methods are similar but use country-specific emission factors and other data, and Tier 3 methods incorporate complex approaches and models of emission sources. Universal application of top-tier methods would significantly reduce uncertainties in reported emissions but would also significantly increase costs. Thus, at a minimum, top-tier methods should be used for the most important greenhouse gas sources in each country.

Recommendation. Annex I countries should develop and implement standardized methods for preparing and publishing inventories that are gridded at spatial and temporal resolutions appropriate for the particular greenhouse gas and source.

Because the atmosphere is not well mixed at country scales, spatially and temporally heterogeneous emissions imply complex variations in the greenhouse gas abundances both at the surface and in the atmospheric column. Independent estimates of national emissions based on tracer-transport models require some prior knowledge of the pattern of emissions. Many Annex I countries are compiling spatially and temporally resolved greenhouse gas emissions, but a standard method for producing gridded measurements does not yet exist. Gridded inventories would provide information at spatial and temporal scales better matched to the dynamics of the atmosphere and thus facilitate comparisons of reported emissions with atmospheric methods. They would be particularly useful for checking reported HFC, PFC, CFC, SF_6, and fossil-fuel CO_2 emissions. The optimal sampling scheme will vary among emissions sources and greenhouse gases and must balance cost and complexity against value.

Independent Estimation of Fossil-Fuel CO_2 Emissions

Independent verification of the self-reported fossil-fuel CO_2 emissions of individual countries will require additional atmospheric measurements and improved tracer-transport estimates of emissions. The density and coverage of measurements would be improved (1) by establishing new stations near cities and other large local sources and in sparsely sampled regions and (2) by deploying a CO_2-sensing satellite. Measurements of radiocarbon (^{14}C) would enable fossil-fuel CO_2 emissions to be separated from non-fossil-fuel sources and sinks. Together, the new measurements and the gridded inventory estimates in Annex I countries would provide the information necessary to reduce errors in the transport models and to overcome the noise from the natural variability of the biosphere. Information derived from all sources could be synthesized in a data assimilation system to produce accurate estimates of anthropogenic CO_2 emissions and trends at national scales. The three recommendations that follow represent critical components of this larger effort that could be deployed within 3 years.

Recommendation. The National Aeronautics and Space Administration (NASA) should build and launch a replacement for the Orbiting Carbon Observatory (OCO).

Most fossil-fuel emissions emanate from large local sources such as cities and power plants. These large sources increase the local CO_2 abundance in the atmosphere by 1-10 parts per million (ppm), a signal that is significantly larger than the signal from natural sources and sinks. NASA's Orbiting Carbon Observatory, which failed on launch in February 2009, would have had the high precision (1-2 ppm) and small sampling area (1.29 × 2.25 km) needed to monitor these large local sources and to attribute their CO_2 emissions to individual countries. No other satellite has its critical combination of high precision, small footprint, readiness, density of cloud-free measurements, and ability to sense CO_2 near the Earth's surface.

The OCO was designed to study natural CO_2 sources and sinks. It would have demonstrated the technology for estimating CO_2 emissions from space but would have had two limitations for a climate treaty. First, with a revisit period of 16 days, it would have sampled only 7-12 percent of the land surface, enabling only a small percentage of large local emissions sources to be monitored. Second, it would have had a 2-year lifetime, providing only baseline data against which to measure future trends. A replacement for OCO launched in the first few years of the coming decade and a subsequent mission at the decade's end should be able to determine if trends in the number and average intensity of CO_2 "domes" over a country's cities and power plants are consistent with reported fossil-fuel emissions. A replacement mission is expected to cost about the same as the original, $278 million.

Recommendation. Extend the international atmospheric sampling network:

• **To research the atmospheric domes of greenhouse gases over a representative sample of large local emitters, such as cities and power plants, and**
• **To fill in underrepresented regions globally, thereby improving national sampling of regional greenhouse gas emissions.**

The atmospheric sampling network, coordinated by the World Meteorological Organization's Global Atmospheric Watch (GAW) and operated by the National Oceanic and Atmospheric Administration (NOAA) and agencies in other countries, comprises approximately 150 stations around the world that measure a host of greenhouse gases from the ground, ocean surface, and air. The stations in the network were purposely located away from large local emitters to minimize contaminating the signal from natural sources and sinks with the signal from fossil-fuel combustion. However, adding ground stations or aircraft to measure emissions from power plants and cities would enable the network to monitor both types of signals. New measurements of relevant trace gases (e.g., greenhouse gases, isotopes of carbon), their biological fluxes, and meteorological variables would be made at locations radiating from the center of each large emitter. This research initiative would yield data needed to calibrate satellite measurements of large local emitters (see previous recommendation), demonstrate an independent capability to monitor large local emitters from ground stations and aircraft, and document long-term shifts in fossil- versus non-fossil-fuel sources in urban and industrial regions. An initial goal could be to deploy instruments at a statistical sample of large emitters (e.g., 5-10 within a research budget of $15 million to $20 million per year) in the United States, but international partners would ideally extend the effort in other countries.

The GAW network is capable of achieving the sub-ppm precision in CO_2 measurements necessary to detect the atmospheric signal from widespread but unconcentrated sources, such as land-use fluxes. It also provides an opportunity to make concurrent measurements of other gases that are needed to derive total carbon emissions from a region. However, huge areas of the planet are not adequately sampled by this network. For example, there are only a few sites in Africa and South America. Expanding the GAW network to observe the variations in greenhouse gas abundances in countries with the highest emissions would greatly improve the independent verification of emissions through tracer-transport modeling. Expanding the network to obtain frequent vertical profiles from aircraft and balloons would constrain atmospheric transport and allow more meaningful comparisons with satellite retrievals of column-averaged CO_2 than ground-based measurements alone. Negotiators should work toward participation in the cooperative network by all major emitting countries and by groups of neighboring smaller countries. Implementing this recommendation will require financial assistance and capacity building to aid the poorest countries that dominate the most undersampled regions.

Recommendation. Extend the capability of the CO_2 sampling network to measure atmospheric ^{14}C.

Estimating fossil-fuel CO_2 emissions from tracer-transport inversion is complicated by poorly understood natural emissions of CO_2 that fluctuate and can be as large as or larger than those from fossil-fuel sources. Adding ^{14}C measurements to the atmospheric sampling stations that measure CO_2 (CO_2 sampling network) would provide an unambiguous means to differentiate between the CO_2 from fossil-fuel and non-fossil-fuel sources because modern organic material contains radiocarbon from cosmic rays and bomb tests, but the ^{14}C in fossil fuels has long since decayed away. It would also provide key measurement constraints to improve tracer-transport inversions. The $^{14}CO_2$ measurements would enable fossil-fuel use to be estimated at subcontinental scales with uncertainties low enough to be useful for verifying self-reported emissions.

The $^{14}CO_2$ measurements could be made at a small incremental cost (~$5 million to $10 million per year for 10,000 measurements, including half in the United States). This initiative could be undertaken by NOAA, which maintains the CO_2 sampling network and has the facilities and expertise to collect and process the samples; by the Department of Energy, which operates a suitable accelerator mass spectrometer at Lawrence Livermore National Laboratory; or by another national laboratory or a university capable of making the measurements at the required precision. International partners could help extend this capability to other countries, providing a more global capability for verifying fossil-fuel emissions.

Implementation of the recommendations to enhance atmospheric sampling and UNFCCC inventories should lead to rapid improvements in monitoring and verification. Rigorous inventories in all countries, added in situ stations, a replacement for OCO, and the $^{14}CO_2$ measurements would increase the number of high-resolution greenhouse gas measurements by orders of magnitude, improve transport models, and significantly reduce errors associated with natural emissions. The loss of any one of these measurement systems would increase the uncertainty for tracer-transport inversion, making the uncertainty in the emissions estimates greater than the 10-year reductions likely required under an international treaty.

Independent Estimation of Fluxes from Land-Use Sources and Sinks

Emissions and removals from land use are highly uncertain both because of uncertainty in the levels of activities such as deforestation or forest planting and because of uncertainty in the emissions per unit of activity. Implementation of the first recommendation below would provide useful estimates of land-use activity levels that could be used to check self-reported values in UNFCCC inventories and also enable more accurate land-use emissions reporting from developing countries. The second recommendation would deliver improved estimates of the emissions per unit of activity.

Recommendation. Establish a standing group to produce a global map of land-use and land cover change at least every 2 years. This will require a commitment to maintaining the continuous availability, in the public domain, of Landsat (or an equivalent satellite) and high-resolution satellite imagery.

Landsat imagery provides an independent check on the activities that create the largest CO_2 emissions from agriculture, forestry, and other land use. Estimates of global land use and land cover must be made often enough to detect important changes, such as forest clearing or planting (e.g., 1-2 years in most forests). Although individual Landsat scenes are publicly available, regular production of a gridded map of the world would allow country-specific information to be extracted on a routine basis. Such a map would enable countries to validate land-use emissions and would provide a basis for improving land-use inventories in developing countries.

The maps could be produced by the U.S. Geological Survey—which has a long history of creating, distributing, and archiving Landsat products—NASA, or a university. For the satellite platforms, either NASA has to keep a successor to the Landsat Data Continuity Mission in its mission queue or another agency will have to maintain the capability. The moderate-resolution (30 m) imagery should be supplemented with statistical samples of high-resolution (1 m) imagery to monitor logging, forest degradation, and certain agricultural practices (e.g., rice cultivation). Such high-resolution data could be obtained either by adding the capability to the Landsat platform or by acquiring unrestricted targeted samples from government or commercial satellites. Without this medium- and high-resolution imagery, we will lose our capability to check the dominant source of agriculture, forestry, and other land-use (AFOLU) CO_2 emissions.

Recommendation. An interagency group, with broad participation from the research community, should undertake a comprehensive review of existing information and design a research program to improve and, where appropriate, implement methods for estimating agriculture, forestry, and other land-use emissions of CO_2, N_2O, and CH_4.

Methods for producing greenhouse gas inventories evolve as more is learned about how to measure and translate activities into emissions. Improvements to U.S. inventory methods could eventually become part of UNFCCC reporting through the established process managed by the IPCC. The most important component to improve is agriculture, forestry, and other land-use emissions of CO_2, N_2O, and CH_4, which have the greatest uncertainties in the national inventories, primarily because of high uncertainty in emission factors. Continued research on the biogeochemical cycles of these gases is needed, especially on the CO_2 emissions caused by deforestation and forest degradation, CH_4 emissions from rice paddies and cattle, and N_2O emissions from fertilizer application. Research is also needed on the natural cycles of CO_2, N_2O, and CH_4 because natural emissions interfere with the detection of anthropogenic signals. This research has to be supported by ecosystem flux observations and ecosystem inventories. For example, eddy covariance towers measure the exchange of carbon between vegetation and the atmosphere at more than 100 sites in the United States. The towers provide valuable information on trends in ecosystem responses to management and climate, and a subset could be maintained to support verification research at relatively low cost (~$100,000 per station per year).

As with fossil fuels, gridded inventories of emissions from agriculture, forestry, and other land use would facilitate cross-checks with other kinds of measurements recommended above. Currently, the only intensive U.S. ecosystem inventory focuses on forests (the U.S. Department of Agriculture's [USDA's] Forest Service Inventory and Analysis program). Making annual measurements of all the major carbon pools and their trends in other ecosystems—including croplands, pastures, and nonforested natural ecosystems—would greatly reduce their emissions uncertainties, which are commonly greater than 100 percent. The cost of a comprehensive ecosystem inventory would likely be substantially less than the cost of USDA's forest inventory ($65 million per year), which includes more field sites (more than 100,000 plots) than are necessary for greenhouse gas monitoring.

IMPLICATIONS FOR AN INTERNATIONAL CLIMATE AGREEMENT

International agreements to limit future greenhouse gas emissions will require that countries be able to monitor and verify emissions as well as removals by sinks. Within a few years of their implementation, the above recommendations would establish rigorous annual national inventories of greenhouse gas emissions as the core of a monitoring and verification system. Procedural verification by an independent international body would be supplemented by independent and transparent checks on fossil-fuel combustion and deforestation, which together are responsible for about three-fourths of all UNFCCC greenhouse gas emissions. Targeted research would ultimately lead to improved monitoring and verification of all greenhouse gases.

Realistic near-term goals are to reduce uncertainties of fossil-fuel CO_2 emissions to less than 10 percent in annual, national inventories and to provide checks on these emissions, especially from large, high-emitting countries—such as the United States, China, or India—using independent methods that are equally accurate. Although national inventories of AFOLU emissions are currently relatively inaccurate, a realistic near-term goal is to reduce uncertainties of AFOLU CO_2 emissions and to be able to estimate remotely the most important activities that cause these emissions (deforestation, afforestation, and forest degradation) with <10 percent uncertainty. In contrast, fundamental research is needed before it will be possible to estimate national emissions of N_2O, CH_4, and the synthetic fluorinated gases with reasonable accuracy using independent methods. The need for fundamental research is especially evident in the high uncertainties for emissions of CH_4 and N_2O from all important AFOLU sources, even for estimates from improved inventory methods.

In addition to improving estimates of AFOLU emissions, the satellite surveys and inventory improvements recommended in this report would allow monitoring of individual projects aimed at creating carbon sinks to offset emissions. The ecosystem inventories would provide the baselines against which an offset project could demonstrate its effect on carbon uptake, which is necessary because carbon fluxes to and from ecosystems fluctuate with the weather and other factors. They would also provide a means for monitoring natural sinks and sources on unmanaged land.

An additional benefit of the proposed expansion of the system to monitor greenhouse gases is that it would enhance our ability to monitor and study natural carbon sinks on land and in the oceans. The natural sinks are not counted in UNFCCC inventories, but they currently absorb about half of greenhouse gas emissions (approximately evenly divided between land and oceans). Because they are so large, changes in the natural sinks could weaken the impact of a treaty. The proposed additions to atmospheric sampling, inventories, and tracer-transport inversion would significantly improve our ability to monitor and study the natural sinks.

1

Introduction

Evidence that climate is changing—including increasing global temperatures, melting glaciers, rising sea level, increasingly severe weather, and shifting seasons and animal migration patterns—is driving national and international discussions on reducing anthropogenic greenhouse gas emissions, the primary cause of climate change. The principal international framework for greenhouse gas reductions is the United Nations Framework Convention on Climate Change (UNFCCC), which is aimed at "stabilization of greenhouse gas concentrations in the atmosphere at a level that would prevent dangerous anthropogenic interference with the climate system" (United Nations, 1992, p. 4). The greenhouse gases covered by the UNFCCC include carbon dioxide (CO_2), methane (CH_4), nitrous oxide (N_2O), sulfur hexafluoride (SF_6), perfluorocarbons (PFCs), and hydrofluorocarbons (HFCs).[1] In 1997, the parties to the UNFCCC approved the Kyoto Protocol, which contains binding emissions targets for developed countries (United Nations, 1998). The United States is not a party to the Kyoto Protocol, but it is considering a variety of proposals for reducing emissions to mitigate adverse effects of climate change, including an international climate treaty.[2]

For any international agreement to limit greenhouse gas emissions, monitoring and verification of emissions will be essential to assess the effectiveness of emissions reductions and overall compliance with the terms of the treaty and to give nations confidence that their neighbors are also living up to their commitments. As former president Ronald Reagan said: "Trust but verify." Emissions verification will also be important for correcting errors in reporting.

DOMAIN OF THE REPORT

This report examines methods for estimating anthropogenic greenhouse gas emissions and for observing their changes over time (see committee charge in Box 1.1). The report asks: How accurate is each method for estimating greenhouse gas emissions? How well can emissions reductions required under a climate treaty be monitored? What new measurement

---

BOX 1.1 Committee Charge

The study will review current methods and propose improved methods for estimating and verifying greenhouse gas emissions at different spatial (e.g., national, regional, global) and temporal (e.g., annual, decadal) scales. The greenhouse gases to be considered are carbon dioxide, chlorofluorocarbons (CFCs), hydrofluorocarbons (HFCs), nitrous oxide, methane, and perfluorinated hydrocarbons (PFCs). Emissions of soot and sulfur compounds along with precursors of tropospheric ozone may also be considered. The results would be useful for a variety of applications, including carbon trading, setting emissions reduction targets, and monitoring and verifying international treaties on climate change.

---

[1] A separate treaty, the Montreal Protocol on Substances That Deplete the Ozone Layer, covers chlorofluorocarbons (CFCs) and hydrochlorofluorocarbons (HCFCs).

[2] International negotiations are intended to culminate in an agreement at a future Conference of the Parties to the UNFCCC, see <http://unfccc.int/2860.php>.

methods could be developed within a few years to independently verify emissions estimates?

The focus of this report is on monitoring and verification of the emissions themselves (see definitions in Box 1.2), rather than on implementation of policies designed to control them. The scales of interest range from national to global and from annual to decades. Although some of the methods described in this report have sufficiently high resolution to be used to audit individual emissions sources, which may be of inter-

---

BOX 1.2 Definitions of Terms Used in the Report

Activity data—Data on the magnitude of a human activity resulting in emissions or removals during a given period of time. Examples include data on energy use, metal production, management systems, forest clearing, and fertilizer use.

Annex I countries—The 41 countries included in Annex I (as amended in 1998) to the UNFCCC, including industrialized countries that were members of the Organisation for Economic Co-operation and Development in 1992 and many countries with economies in transition. Under the convention, Annex I countries committed to returning individually or jointly to their 1990 levels of greenhouse gas emissions by 2000. By default, the other countries are referred to as non-Annex I countries.

Anthropogenic emissions—Emissions of greenhouse gases, precursors of greenhouse gases, and aerosols resulting from human activities. Because it is difficult to disentangle anthropogenic and natural components of emissions and removals from land use, the UNFCCC considers emissions and removals on managed lands as anthropogenic.

CO_2 equivalent—The amount of carbon dioxide emission that would cause the same integrated radiative forcing, over a given time horizon, as an emitted amount of a well-mixed greenhouse gas. It is a standard metric for comparing emissions of different greenhouse gases, but does not imply exact equivalence of the corresponding climate change responses. The 100-year global warming potential is used to calculate CO_2 equivalents.

Emission factor—The rate of emission per unit of activity, output, or input. For example, a particular fossil-fuel power plant may have a CO_2 emission factor of 0.765 kg CO_2 kWh^{-1} generated.

Inventory—An accounting of an item of interest at a specified date.
- An emissions inventory accounts for the amount of one or more specific greenhouse gases discharged into the atmosphere from all source categories as well as removals by sinks in a certain geographical area and within a specified time span, usually a specific year. Under the UNFCCC, Annex I countries prepare national inventories of anthropogenic greenhouse gas emissions and removals for each calendar year.
- An ecosystem inventory accounts for the carbon stored in a particular land classification (e.g., forest, peatland) based on ecosystem characteristics that affect carbon storage, such as volume of soil carbon and live and dead above- and belowground biomass, measured from a network of plots. Changes in carbon stock through time (i.e., carbon uptake or release) are measured by differencing two samples from the same plot but separated by 1 to 10 years.

Inverse model—A model in which observations are used to infer the values of the parameters characterizing the system under investigation. In this report, inverse models are used to infer sources and sinks for a greenhouse gas from measurements of the atmospheric or oceanic abundance of that gas.

Monitoring—The observation of emissions or variables correlated with emissions for the purpose of detecting any changes that may occur over time.

Sector—An emission-producing segment of the economy. The Intergovernmental Panel on Climate Change (IPCC) currently specifies four sectors for greenhouse gas reporting: energy; industrial processes and product use; agriculture, forestry, and other land use; and waste.

Sink—Any process, activity, or mechanism that removes a greenhouse gas, an aerosol, or a precursor of a greenhouse gas or aerosol from the atmosphere. Removals of greenhouse gases by a sink are conventionally shown as negative emissions.

Source—Any process, activity, or mechanism that releases a greenhouse gas, an aerosol, or a precursor of a greenhouse gas or aerosol into the atmosphere. Certain activities, such as forestry, can be both a source and a sink of greenhouse gas emissions.

Survey data—Data from a statistically representative sample.

Tracer-transport model—A model used to predict the movement of greenhouse gases in the atmosphere or dissolved substances in the oceans.

Verification—An independent examination of monitoring data to help establish whether or not a country's actual emissions are consistent with its obligations under a climate treaty.

SOURCES: Adapted from IPCC glossaries (<http://www.ipcc.ch/>) and UNFCCC resources (<http://unfccc.int/2860.php>).

est for trading schemes or offset projects, the report focuses on the national emission totals that include these activities. Only public domain data (not classified or commercial data) are considered because confidence in a treaty relies on open data for transparency and scientific scrutiny.

Greenhouse Gases

This report considers the anthropogenic greenhouse gases required by the committee charge—CO_2, CH_4, N_2O, CFCs, HFCs, and PFCs—and SF_6, but not the optional soot or precursors of tropospheric ozone. The greenhouse gases required by the committee charge, along with SF_6, are currently covered by international agreements (CFCs under the Montreal Protocol and the others under the UNFCCC) and were the targets of negotiations at the 2009 United Nations Climate Change Conference (COP 15) in Copenhagen. Thus, there is an immediate practical need to verify emissions of the gases included in this report, which does not extend to the greenhouse agents that were omitted. The short-lived greenhouse agents (soot and other aerosols, aerosol precursors, and precursors to tropospheric ozone) are not covered by international agreements, although many countries have a highly developed capability to monitor them to support air pollution regulations. A comparable capability for the greenhouse gases discussed in this report does not exist.

The focus of international agreements on CO_2, CH_4, N_2O, CFCs, HFCs, PFCs, and SF_6 is likely to continue for three reasons. First, these gases are collectively more important greenhouse agents than soot, sulfur compounds, and precursors of tropospheric ozone (Figure 1.1). Commonly cited mitigation targets, such as a maximum of 2°C of warming or a maximum concentration of 450 parts per million (ppm) CO_2 equivalent, cannot be achieved without large reductions in emissions of CO_2, CH_4, N_2O, CFCs, HFCs, PFCs, and SF_6. Second, the gases included in the report are long-lived in the atmosphere (decades to millennia or more), whereas the omitted gases and soot are short-lived (less than a year). Longevity in the atmosphere means that delayed mitigation is costly—CO_2 emissions today will add to global climate change for centuries. In contrast, short-lived greenhouse agents

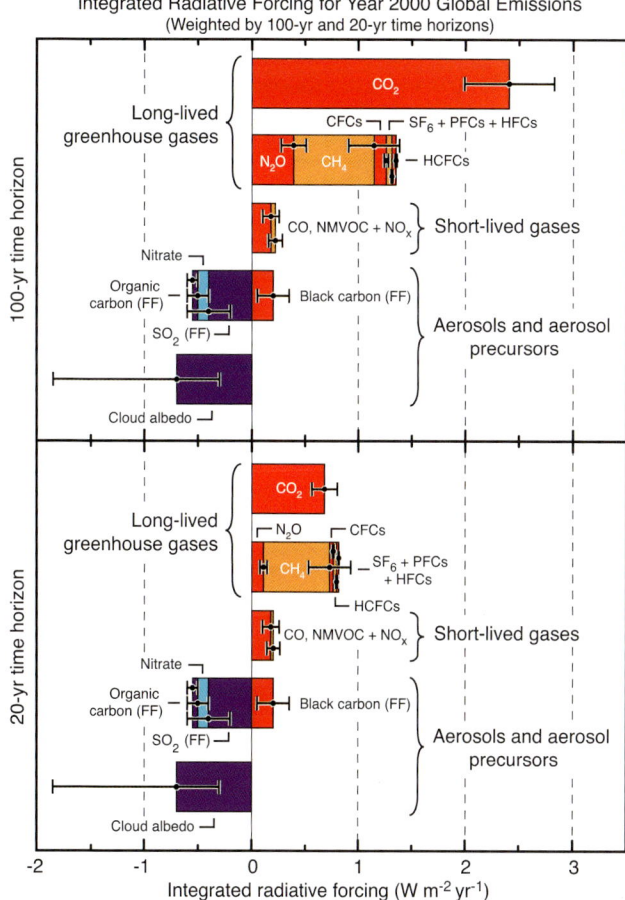

FIGURE 1.1 The relative importance of emissions of anthropogenic greenhouse gases and soot (black carbon) and other aerosols. The bars show the 20- (lower panel) and 100-year (upper panel) radiative forcing of emissions in 2000. SOURCE: Figure 2.22 from IPCC (2007a), Cambridge University Press.

do not entail the same penalties for delay. Because they are removed from the atmosphere in less than a year, today's emissions will have a smaller impact on global warming in coming decades when the problem becomes most acute. Third, the net radiative forcing from the emission of short-lived gases and aerosols depends greatly on the location and timing of emissions. The time required for air to mix globally is on the order of 2 weeks in the east-west direction and 1 year in the north-south direction across the equator, which is less than the lifetime of short-lived greenhouse agents. For this and other reasons, the greenhouse impact of the ozone precursor NO_x (nitrogen oxide) can vary by a factor of 10, depending on whether it is emitted in northern Europe or in the tropics (Wild et al., 2001; see also Table 2.15 of Forster et al., 2007). This makes it difficult to design a practical international agreement to monitor them as greenhouse agents.

Of the long-lived greenhouse gases included under the UNFCCC, CO_2 is responsible for 77 percent of the greenhouse forcing on a 100-year time horizon (Figure 1.2). For this reason, the report devotes considerably more space to CO_2 than to the other gases. Note, however, that if we include the short-lived gases and soot and adopt a 20-year horizon, the contribution of CO_2 falls to less than 40 percent.

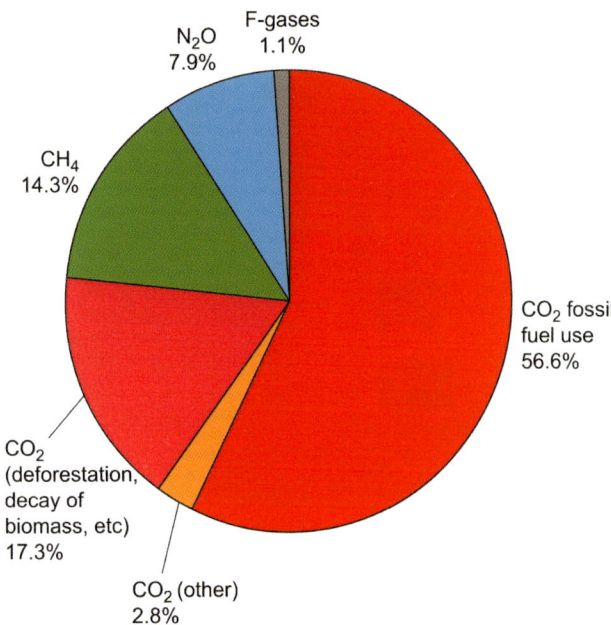

FIGURE 1.2 Global anthropogenic greenhouse gas emissions in 2004. The F-gases include HFCs, PFCs, and SF_6. The area in the pie diagram shows 2004 emissions of the gases covered by the UNFCCC, weighted by their 100-year radiative forcing. SOURCE: Figure 1.1b from IPCC (2007b), Cambridge University Press.

Methods of Monitoring

This report evaluates three categories of monitoring approaches: national inventories, satellite measurements of land use, and atmospheric methods.

National Inventories. Under the UNFCCC, Annex I (developed) countries are required to report annual anthropogenic emissions and removals of greenhouse gases. Developing countries also report national inventories, but less frequently and in far less detail than developed countries. The emissions estimates are based on measurements of human activities (i.e., data such as cement or coal-fired electricity production) and corresponding emission factors (see definitions in Box 1.2). Because future international agreements are likely to build on this foundation, the committee evaluates UNFCCC inventory methods and extensions of them that would improve their comprehensiveness and accuracy and increase the rigor of self-reporting. The committee also discusses the capacity building necessary to procure regular inventories from developing countries.

Remote Sensing of Land Use. Greenhouse gas emissions and sequestration from land use are difficult to estimate because they have the same chemical signature as much larger background sources and sinks in the natural biosphere and because they are thinly spread over an enormous area. The dominant sources of land-use emissions are from forestry (primarily tropical deforestation and forest degradation) and agriculture. Land-use emissions in parts of the temperate zone are negative (i.e., net removals by sinks) due to net forest regrowth and other processes (Pacala et al., 2007). Because deforestation is the second largest source of anthropogenic CO_2 (the first is fossil-fuel combustion) and because forest conservation and planting are likely to be important mitigation activities in the future, this report devotes considerable attention to methods for monitoring forest cover and structure by satellites. A comparable understanding of N_2O and CH_4 emissions from croplands and grasslands does not exist, both because of the diversity of agricultural practices and because we lack the technology to measure the dominant emissions sources remotely. For these reasons, it remains difficult to provide a useful check on self-reporting of emissions from agriculture, except in specific instances.

Atmospheric Methods. A global network of surface monitoring stations, aircraft, balloons, and satellites routinely measures greenhouse gas abundances in the atmosphere and oceans. Models of the atmosphere and/or oceans are used to estimate greenhouse gas emissions from the abundance data, a method known as tracer-transport inversion. An emissions source located between two monitoring stations will cause the concentration of the gas to be higher at the downwind station than the upwind station. How much higher depends on both the strength of the source and the pattern of air flow, including wind speed, direction, and turbulence. Thus, to produce emissions estimates from abundance data, one needs an atmospheric model to reconstruct the three-dimensional pattern of air and water flow and mixing around the globe. For this reason, the report devotes considerable space to uncertainties in atmospheric transport models.

The report also evaluates extensions of the atmospheric sampling network that could significantly improve our ability to estimate national emissions and emissions trends. These include measurements of concentrations that would fill spatial gaps in the current sampling grid—for example, samples taken near large sources such as power plants and municipalities that were avoided when the current sampling network was established.

Uncertainty

This report evaluates uncertainties in annual emissions estimates derived from the three monitoring methods described above. In some cases, standard statistical methods can be used to evaluate the uncertainties, but in others, standard methods cannot be applied because our underlying scientific understanding is too incomplete or our measurement capabilities are insufficient. In such cases, we rely on other methods, including expert judgment, that are specified in tables of uncertainty estimates. Uncertainties are categorized in five bins—0-10 percent, 10-25 percent, 25-50 percent, 50-100 percent, and >100 percent (for the last category, it is unclear whether the activity is a source or a sink)—to facilitate cross-comparison between estimates from different methods. An uncertainty of 10 percent means that measurements are accurate to within 10 percent of the true value. Unless indicated otherwise, uncertainties are reported for two standard deviations of the mean (2σ or 95 percent confidence interval).

Uncertainties in decadal changes can be computed from the values for annual emissions using standard time-series methods, including simple regression, for calculating the uncertainty of regression slopes. A reasonable expectation is that uncertainties in the decadal change of emissions will be lower than the annual uncertainty.

OVERVIEW OF GREENHOUSE GAS EMISSIONS

Relative Contribution to Climate Change

A greenhouse gas's instantaneous tendency to change the climate is measured by its radiative forcing, which multiplies the increased abundance of the gas caused by anthropogenic emissions and the gas's potency as a greenhouse agent. Of the four groups of gases considered in this report, CO_2 has the largest radiative forcing (1.66 W m^{-2} for emissions up to the end of 2005), followed by CH_4 (0.48 W m^{-2}), the HCFCs and CFCs (collectively 0.32 W m^{-2}), N_2O (0.16 W m^{-2}), and the HFCs, PFCs, and SF_6 (collectively 0.02 W m^{-2}; see Forster et al., 2007).

Longevity in the Atmosphere

The longevity of a greenhouse gas in the atmosphere is important because it determines the number of years that today's emissions will affect climate. Short-lived gases, such as the precursors to tropospheric ozone, are rapidly cleared from the atmosphere; thus, the perturbation caused by emissions appears to adjust rapidly to a change in emissions. However, short-lived chemically reactive gases are coupled with the longer-lived greenhouse gases and thus produce long-lived perturbations to radiative forcing that take decades to reach a steady state (Wild et al., 2001). For even small levels of anthropogenic emissions, the atmospheric abundances of very long-lived gases, such as the PFCs, will continue to rise in proportion to emissions and remain well below the steady-state value at which annual emissions are balanced by annual removals. To eventually halt climate changes caused by greenhouse gases, their abundances in the atmosphere must be stabilized.

The lifetime of CO_2 in the atmosphere cannot be ascribed a single value because the carbon cycle consists of a series of interacting reservoirs, each with a different time scale (see Figure 1.3). For example, although land ecosystems or the oceans take up approximately one-sixth of the CO_2 in the atmosphere every year, they also return almost the same amount (IPCC, 2007a). Thus, the lifetime of a pulse increase in the atmospheric abundance of CO_2 is set not by the short stay of an

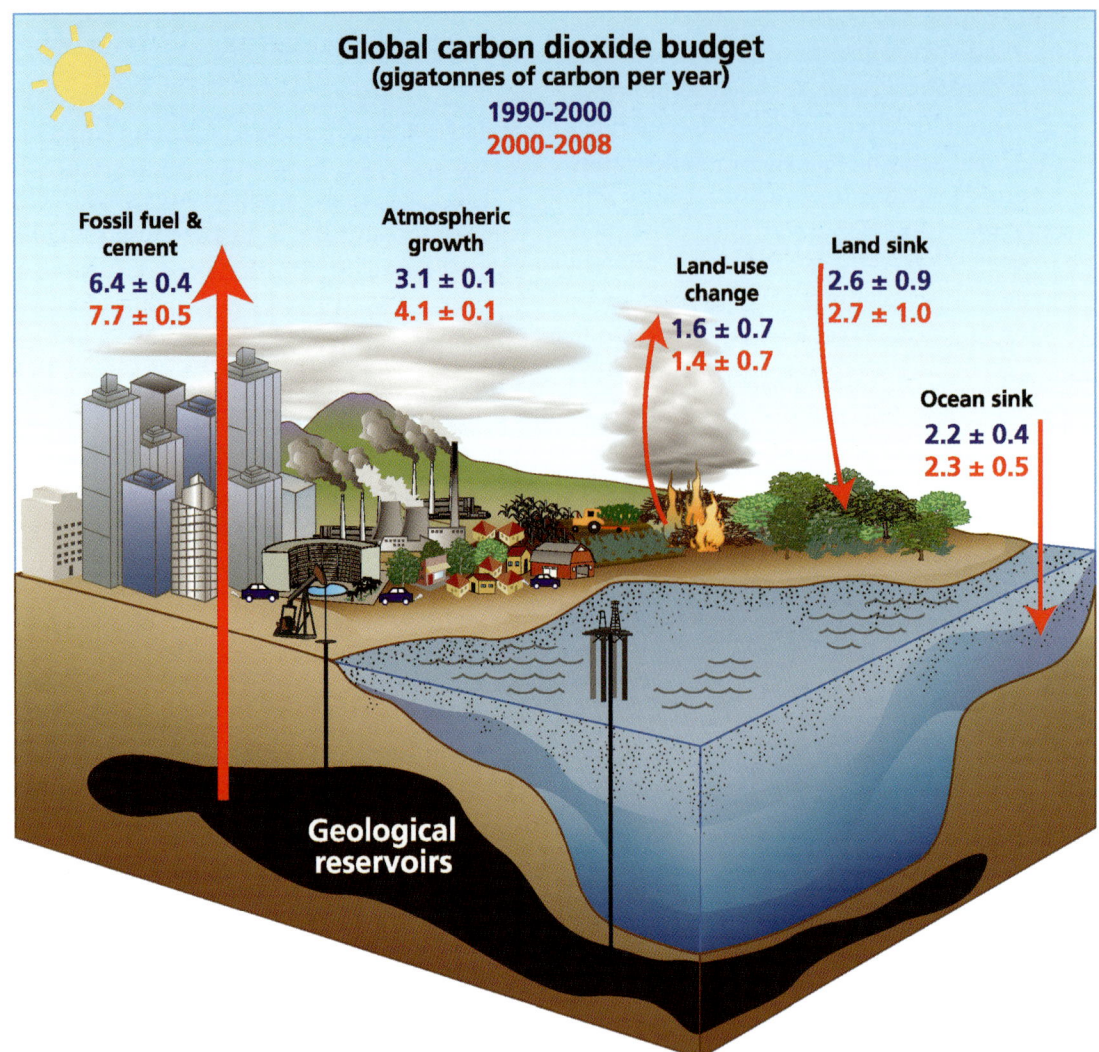

FIGURE 1.3 The global carbon cycle and changes in the sizes of CO_2 reservoirs over the last two decades. All values are billions of metric tons of carbon. Arrows show annual fluxes. Values for the 1990s are in blue; those for 2000-2008 are in red. SOURCE: Le Quéré (2009), International Geosphere-Biosphere Programme/Global Carbon Project. Data from Le Quéré et al. (2009).

individual molecule in the atmosphere, but by the small imbalance that the pulse creates between the uptake and removal rates. For example, CO_2 molecules spend thousands of years in the oceans once they have been transported into the abyss. Consequently, dissolved CO_2 in the deep oceans reflects the atmospheric abundance before the industrial revolution, rather than the increased abundance caused by fossil-fuel burning in the last 200 years. The upshot is that a fraction of the fossil-fuel CO_2 emitted is taken up rapidly by the upper ocean and biosphere, but the remainder of the perturbation acts like a very long-lived gas, requiring thousands of years to decay away (Denman et al., 2007).

Methane is short-lived in the atmosphere relative to CO_2. A molecule emitted into the atmosphere is oxidized to CO_2 in an average of about 8 years, but chemical feedbacks extend this time scale to 12 years (Prather, 1994). This means that the current abundance of methane is derived from the last several decades of emissions. Nitrous oxide has an average residence time of 114 years in the atmosphere before it is photochemically decomposed in the stratosphere. The average atmospheric lifetimes of the other gases considered in this report range from 45 to 1,700 years for CFCs, 1 to 270 years for HFCs, 3,200 years for SF_6, and tens of thousands of years for PFCs (Forster et al., 2007).

Emission Sources

CO_2 emissions to the atmosphere are caused primarily by fossil-fuel burning (~74 percent in 2004; IPCC, 2007b) and tropical deforestation (~22 percent), although recent work suggests that the contribution from deforestation has decreased to as little as 12 percent of CO_2 emissions in 2008 (van der Werf et al., 2009a). Other contributors include industrial processes such as cement production. The primary sources of anthropogenic methane are energy production, ruminant animals, rice agriculture, landfills, and biomass burning (Denman et al., 2007). Natural sources of methane are dominated by wetlands and are approximately one-half the size of anthropogenic sources. Although understanding of anthropogenic N_2O sources is incomplete, agriculture is likely the largest source because of the oxidation of nitrogen fertilizer and reduction of nitrite (see Table 7.7 in IPCC, 2007a). Natural sources are dominated by soils under natural vegetation and by microbial transformations of nitrogen compounds in the oceans and are thought to be roughly comparable in size to anthropogenic sources (Boumans et al., 2002; Nevison et al., 2004; Hirsch et al., 2006). The presence of CFCs, HFCs, SF_6, and PFCs in the atmosphere is due almost entirely to human manufacture for a wide range of industrial applications in the latter half of the twentieth century (IPCC, 2007a).

Atmospheric Concentrations, Emissions, and Trends

Atmospheric abundances of greenhouse gases are best quantified by dry air mole fractions—the number of molecules of the gas in a set volume divided by the total number of molecules of dry air in the same volume (see Box 1.3). The mole fraction of CO_2 in the atmosphere is currently 387 ppm (Figure 1.4), which is more than 100 ppm higher than in the pre-industrial period. Annual anthropogenic emissions of CO_2 are between 9 billion and 10 billion metric tons of carbon (Gt C yr^{-1}), increased at 1-2 percent per year over the last three decades of the twentieth century, and 3.4 percent per year from 2000 to 2008, and are projected to decline in 2009 by almost 3 percent due to the weak economy (Canadell et al., 2007; Le Quéré et al., 2009). Approximately half of the annual increase expected from these emissions (4-5 Gt C yr^{-1}) accumulates in the atmosphere and the rest is taken up by carbon reservoirs in the oceans and on land (see discussions below and Figure 1.3). Because 1 ppm of CO_2 in the atmosphere equals 2.12 Gt C, the atmospheric growth rate currently averages ~2 ppm per year.

The abundance of methane in the atmosphere today is much higher than in the millennium before the industrial era (1,774 parts per billion [ppb] in 2005 versus approximately 700 ppb; see IPCC, 2007a). For the 1970s and 1980s, the growth rate of methane was about 1 percent per year; the rate slowed dramatically in the 1990s and dropped to nearly zero from 2000, but began to grow again in 2007 (Rigby et al., 2008b).[3] Several reasons for this anomalous growth pattern have been proposed, but no clear explanation is available (Dlugokencky et al., 2001). The literature on emissions of methane is summarized in Table 7.6 of IPCC (2007a). Estimates for anthropogenic sources range from 264 to 428 Tg CH_4 yr^{-1} (1 Tg CH_4 equals one million metric tons of methane) and for natural sources from 145 to 260 Tg CH_4 yr^{-1}, although total emissions are more tightly constrained (493 to 667 Tg CH_4 yr^{-1}).

---

[3] See also <http://www.noaanews.noaa.gov/stories2008/20080423_methane.html>.

> **BOX 1.3 Measurement Units for Greenhouse Gases in the Atmosphere**
>
> The concentration of a gas, which is defined as the number of molecules per volume, will vary with altitude and weather systems as the density of the air changes, even if there are no sources or sinks. When a parcel of air rises and expands at lower pressure, the concentrations of all species decrease by the same factor. What is conserved is the mole fraction, the relative abundance of each. When water evaporates or condenses, which adds or removes an extra gaseous component, the mole fraction of all other components will decrease or increase, respectively, by the same proportion. Thus, the property that reflects additions and removals of a trace component is its mole fraction in dry air, which changes only when there are sources or sinks. The dry air mole fraction of CO_2 is expressed as parts per million. A mole fraction of 385 ppm means that, on average, in every 1 million molecules of dry air there are 385 CO_2 molecules. The mole fraction of methane is typically expressed in parts per billion, and that of the HFCs, PFCs, and CFCs in parts per trillion.

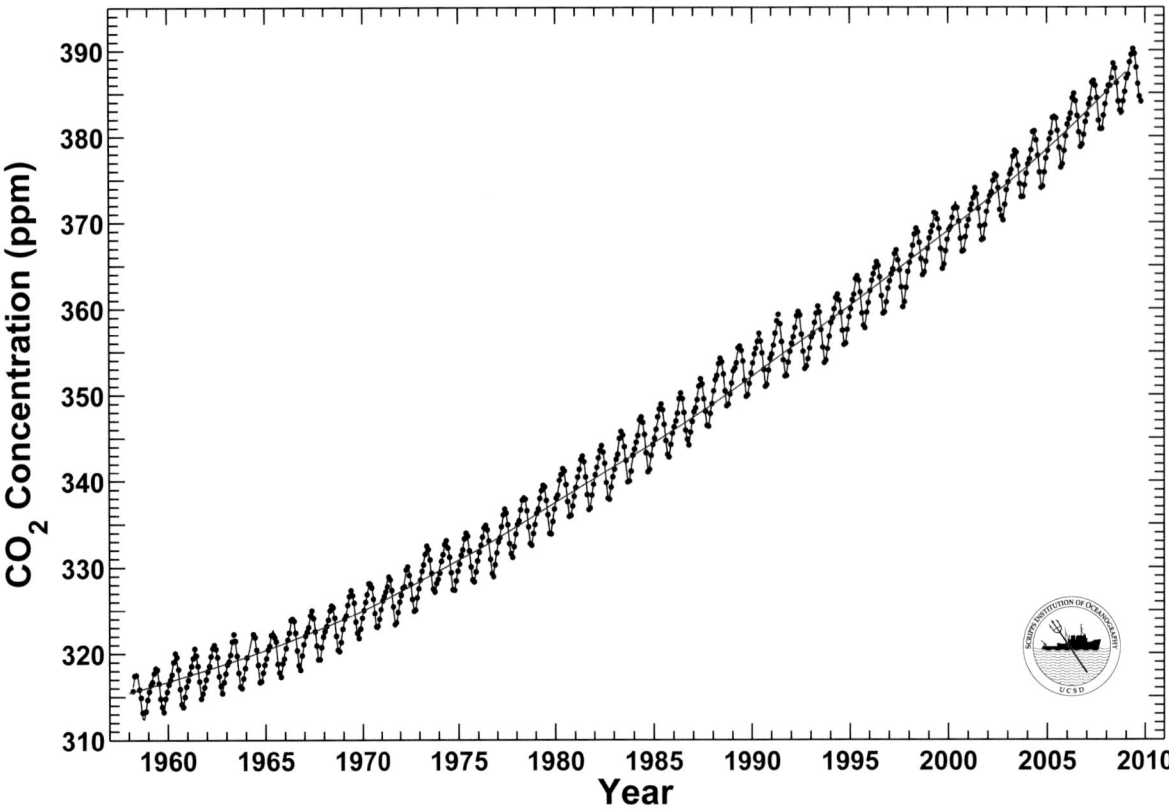

FIGURE 1.4 Monthly mean CO_2 concentration at 3,400 m altitude on Mauna Loa, Hawaii. The red curve shows the trend of industrial emissions of CO_2 from fossil-fuel combustion and cement production. The annual cycle is driven by the imbalance between seasonal photosynthesis and respiration on the continents. Plants take up CO_2 only during the growing season, but plants and animals release it through plant metabolism and the decay of dead organic matter more evenly throughout the year. The long-term increase in atmospheric CO_2 is caused by fossil-fuel combustion and land-use change. SOURCE: Courtesy of Ralph Keeling, Scripps Institution of Oceanography. Data are from the Scripps CO_2 program.

The abundance of N_2O rose from 270 ppb in 1750 to 319 ppb in 2005 (IPCC, 2007a). N_2O is now increasing in concentration at an average rate of approximately 0.25 percent per year. Total emissions (13.9 to 18.9 Tg N yr^{-1}) are constrained by the predicted atmospheric lifetime and observed growth rate (Prather et al., 2001), with the growth rate indicating the level of anthropogenic emissions (~6 Tg N yr^{-1}). Inventory estimates of N_2O emissions have considerable uncertainties (Prather et al., 2009) as illustrated by the large range in the size of the total global source reported by the Intergovernmental Panel on Climate Change (IPCC): 8.5-27.7 Tg N yr^{-1} (IPCC, 2007a, Table 7.7).

CFCs are covered by the Montreal Protocol on Substances That Deplete the Ozone Layer, and they are either stabilized or decreasing in concentration. However, HFC, SF_6, and PFC abundances are currently increasing (Prinn et al., 2005; Velders et al., 2005; IPCC, 2007a).

Separating Anthropogenic and Natural Components of CO_2

Anthropogenic emissions are the emissions of a gas resulting from human activities. Although this definition is easy to apply to fossil-fuel burning, where all emissions are anthropogenic, it becomes problematic for forestry, cropland management, and other land-use sources and sinks, where it is difficult to distinguish emissions and removals due to human influence (e.g., management practices) from those due to natural factors. For example, climate change and fertilization of plants by anthropogenic CO_2 and nitrogen deposition probably affect plant growth rates all over the world, but our understanding of these effects is incomplete and will likely remain so for the foreseeable future. To address the problem, the IPCC and UNFCCC have adopted a convention of treating all emissions and removals (sinks) on land that is managed, as anthropogenic.[4] Any changes in emissions and removals from these lands are thus considered anthropogenic, regardless of whether natural factors contributed to those changes.

In addition to the definitional ambiguities, monitoring of anthropogenic CO_2 emissions is greatly complicated by the natural cycling of CO_2 through the terrestrial biosphere and oceans (Figure 1.3). The terrestrial biosphere takes up approximately 120 Gt C yr^{-1} through photosynthesis and releases almost all of it back to the atmosphere through respiration by plants, animals, and microbes (IPCC, 2007a). Photosynthesis occurs only during daylight hours in the growing season, whereas respiration occurs at all times, albeit at a reduced rate in some seasons (i.e., winter outside the tropics). This diurnal and seasonal imbalance can be quite large; the CO_2 sources and sinks that they create are often larger than fossil-fuel fluxes in the same location, except in cities or close to power plants where fossil-fuel emissions are concentrated. Moreover, if we ignore tropical deforestation, terrestrial ecosystems represent a net sink that averaged 2.7 ± 1.0 Gt C yr^{-1} over 2000-2008 (Le Quéré et al., 2009). The cause of this sink is not completely understood, although a substantial fraction is due to forest regrowth and other land-use changes in the temperate zone (CCSP, 2007, Chapters 2 and 3) and the remaining fraction may be caused by CO_2 fertilization (Friedlingstein et al., 2006). The size of the net terrestrial flux can change from year to year by as much as 5 Gt C (Baker et al., 2006a), in part from anthropogenic fires in tropical forests associated with El Niño events (Randerson et al., 2005; van der Werf et al., 2009b), but is usually within a range of ±1 Gt C yr^{-1}.

The oceans are also a sink for carbon averaging 2.3 ± 0.5 Gt C yr^{-1} from 2000 to 2008 (Le Quéré et al., 2009). By measuring the changing chemical properties (e.g., pH, pCO_2) of the surface ocean from research vessels and commercial ships of opportunity, the annual sink assignable to an ocean basin can be estimated to a precision of about ±10 percent (e.g., Watson et al., 2009). These measurements show that variations in the oceanic sink are too small to explain the multi-gigaton fluctuations in the atmospheric increase of CO_2 (IPCC 2007a; Le Quéré et al., 2009). Moreover, because of their comparatively high accuracy, estimates of the oceanic sink provide a valuable constraint on estimates of the magnitude of land sinks at regional and global scales (because the land sink equals the fossil-fuel source minus ocean uptake minus the atmospheric increase).

Fluctuations of natural CO_2 sources and sinks create a difficult signal-to-noise problem for efforts to estimate anthropogenic emissions with atmospheric measurements. Seasonally fluctuating background sources and sinks that contribute to the CO_2 signal may be of the same order as the emission reductions that might be required under a treaty. The signal-to-noise problem is further exacerbated by the fact that annual fossil-fuel and deforestation emissions represent only about 1 percent of the CO_2 in the atmosphere (IPCC, 2007a). This means that anthropogenic emissions will change the average CO_2 abundance by only a small amount as air moves across a country over a period of hours to a few days. Thus, an effective way to uniquely identify many large emissions sources is to measure the perturbation in air close to the source, before mixing dilutes the added CO_2. The plume of increased concentration above a major point source can be of order 1-10 percent above the background concentration (see Chapter 4).

Because of the signal-to-noise problem, the natural carbon cycle would have to be monitored as part of any effort to monitor anthropogenic emissions. Monitoring the carbon cycle would also constrain estimates of "leakage," in which reduced emissions in one region or sector lead to increased emissions in another (i.e., soil carbon releases from land newly cultivated for biofuels; see Searchinger et al., 2008; Tilman et al., 2009). Further, the effectiveness of any climate treaty is based on the stabilization of greenhouse gas abundances, whether from anthropogenic or natural sources. Current models indicate that climate change feeds back on natural ecosystems and the ocean to produce new sources or reduce sinks of greenhouse gases, with most of the feedbacks amplifying climate change. For example, warming might cause arctic tundra to emit large quantities of CO_2 and CH_4, causing further climate change, even more releases of CO_2 and CH_4, and so on in a positive feedback loop (Walter et al., 2006; Zimov et al., 2006; IPCC, 2007a; Schuur et al., 2009). These effects need to be detected early to ensure that

---

[4] This rule is not applied consistently, however, because individual countries may define what constitutes managed lands for their national inventories.

any agreement to limit anthropogenic emissions has the desired outcome.

ORGANIZATION OF THE REPORT

This report examines methods used to estimate greenhouse gas emissions and identifies enhancements or new techniques that could be used to significantly improve emissions estimates over the next few years. Chapter 2 describes the national greenhouse gas inventories reported under the UNFCCC and their limitations. The chapter focuses on the accuracy of estimates for the gases and activities in the sectors responsible for most of the emissions: energy and agriculture, forestry, and other land use. The primary gases in these two sectors are CO_2, CH_4, and N_2O. Corresponding estimates for gases and activities in the industrial processes and waste sectors are summarized in Appendix A. The primary gases in these sectors are CO_2, CH_4, N_2O, and HFCs. Chapter 3 describes remote sensing measurements of land use that could provide independent estimates of deforestation and some types of agricultural production. It also describes observations and research needed to improve UNFCCC inventories of emissions from agriculture, forestry, and other land use. Chapter 4 examines atmospheric-based estimates of greenhouse gas emissions, which could provide independent checks on emissions from fossil-fuel use and industrial processes. Additional information on sources of atmospheric and oceanic data, methods for estimating atmospheric signals, and technologies for measuring emissions from large local sources appears in Appendixes B, C, and D, respectively. Biographical sketches of committee members appear in Appendix E, and a list of acronyms and abbreviations is given in Appendix F.

2

National Inventories of Greenhouse Gas Emissions

All countries that are party to the United Nations Framework Convention on Climate Change (UNFCCC) are required to provide national inventories of emissions and removals of greenhouse gases due to human activities. These inventories form the basis for monitoring the progress of individual countries in reducing emissions and for assessing the collective effort of countries to mitigate climate change. The inventories provide self-reported estimates of selected anthropogenic greenhouse gases for four sectors: energy; industrial processes and product use; agriculture, forestry, and other land use (AFOLU); and waste. Countries prepare the estimates using methods developed by the Intergovernmental Panel on Climate Change (IPCC) and approved by the UNFCCC. The methods generally involve multiplying national data on an emissions-generating activity, such as cement production, by an emission factor that specifies greenhouse gas emissions per unit of activity. This chapter describes current practices for developing greenhouse gas inventories, summarizes the accuracy of estimates for the gases and activities responsible for most of the emissions, and identifies improvements that would facilitate monitoring of emissions for an international climate treaty.

DEVELOPING AND REPORTING NATIONAL INVENTORIES

UNFCCC National Inventory Reporting and Review

UNFCCC reporting and review requirements for national inventories differ for developed (Annex I) and developing (non-Annex I) countries. As a result, the scope and quality of national inventories vary greatly. Developed countries annually report calendar-year estimates for all sources and sinks of the six greenhouse gases specified by the UNFCCC (carbon dioxide [CO_2], methane [CH_4], nitrous oxide [N_2O], sulfur hexafluoride [SF_6], perfluorocarbons [PFCs], and hydrofluorocarbons [HFCs]) going back to 1990. The estimates are broken down by sector and into categories within a sector (e.g., aluminum production within the industrial sector). The national inventories, along with detailed documentation of the methods and data sources used to calculate emissions and removals, are submitted electronically in a standard format to facilitate data analysis and comparison. Similar regulations govern reporting of chlorofluorocarbons (CFCs), but these fall under agreements other than the UNFCCC.

The national inventories of developed countries are subject to international review by teams of greenhouse gas inventory experts. These reviews do not attempt to reconstruct the inventory or verify estimates with independent data, but rather assess whether correct methods and appropriate data sources were used to produce the inventory. Statistical analysis of reported data is also performed to identify inconsistencies within a report or with previously submitted reports. In addition, data are analyzed across countries to determine a range of expected levels of emissions per unit of output or activity (implied emission factors) and to identify deviations from these values. Where possible, data submitted by countries are compared with data compiled by international organizations. For example, national statistics used to estimate energy emissions are compared to International Energy Agency (IEA) data, and statistics used to estimate agriculture emissions are compared to UN Food and Agriculture Organization (FAO) data. If anomalies are identified by this analysis, the review teams dig deeper into that country's methods and data. Unless otherwise indicated, the inventory methods discussed in this chapter pertain to developed countries.

Reporting requirements are much less rigorous for developing countries. Emission inventories are reported only periodically in conjunction with a broader national report of climate change programs and activities. There is no set frequency for these national reports and their submission often depends on the provision of international funding. As a result, most developing countries have submitted only one national inventory to date. Reporting of only CO_2, CH_4, and N_2O is required and only at the sector level, not for categories within each sector. Developing countries are not required to provide emissions trends over time or to document methods and data sources, and their inventories are not reviewed.

IPCC Methodologies

The IPCC's National Greenhouse Gas Inventory Program is responsible for developing methods for creating national inventories of greenhouse gas emissions. The IPCC guidelines describe how to estimate national emissions of CO_2, CH_4, N_2O, SF_6, PFCs, and HFCs from anthropogenic sources and sinks using national statistics (activity data) and activity-based emission factors for the four sectors. Guidance is also provided on data sources, data collection methods, quantification of uncertainties, management of inventories, quality assurance and control, documentation, and data archiving. The guidelines have evolved over time to include more emissions sources and to improve and standardize the methodologies. The first edition of the IPCC guidelines was completed and approved in 1994; the most recent (2006) edition has not yet been endorsed by the UNFCCC and is thus not yet used for reporting purposes. However, the 2006 guidelines are expected to be adopted as the basis for reporting national inventories beginning in 2015.

The IPCC methodologies are intended to yield national greenhouse gas inventories that are transparent, complete, accurate, consistent over time, and comparable across countries. Because different countries have different capacities to produce inventories, the guidelines lay out tiers of methods (typically three) for each emissions source, with higher tiers (Tier 3 is normally the highest) being more complex and/or resource intensive than lower tiers. The higher-tier methods usually incorporate country-specific conditions, data, and emission factors and are thus considered more accurate than the lower-tier methods. For example, the Tier 1 method for calculating CO_2 emissions from stationary combustion uses default emission factors for each fuel type, whereas the Tier 2 method requires each country to develop and use country-specific emission factors for each fuel type (see detailed guidance in Gómez et al., 2006). The Tier 3 method uses emission factors that are not only country-specific, but also differentiated by technology and operating conditions. The choice of method used for a particular source in a particular country depends on (1) the importance of that source to the level and trend of emissions in that country and (2) the resources available to prepare the inventory. Countries are encouraged to use country-specific data and emission factors to the extent possible. However, they are not expected to use higher-tier methods if doing so would jeopardize their ability to estimate other important emissions sources. The scope of the effort to prepare the U.S. inventory is described in Box 2.1.

Implications for Monitoring and Verification

Although multiple greenhouse gases are emitted from multiple activities in multiple sectors, the monitoring and verification problem is comparatively simple because only a few activities and greenhouse gases are responsible for the large majority of emissions. Table 2.1 summarizes emissions by sector for Annex I countries as a group, and Figure 2.1 compares emissions across sectors for Annex I and non-Annex I countries. The most important simplifying message is that well over 90 percent of global greenhouse gas emissions are in the energy and AFOLU sectors, making these sectors an obvious focus for monitoring. Energy alone is responsible for almost 90 percent of total net greenhouse gas emissions from Annex I countries and more than 40 percent of net emissions from developing countries. In both groups, CO_2 from fossil-fuel

> **BOX 2.1 Preparation of the U.S. Greenhouse Gas Emissions Inventory**
>
> Preparation of a national greenhouse gas emissions inventory is a major undertaking, involving many people and agencies in a typical Annex I country. The current version of the IPCC guidelines for inventory preparation is a five-volume set of books with several hundred pages of formulas, supporting data, sample calculations, reporting formats, and technical references. In the United States, the Environmental Protection Agency is ultimately responsible for planning and preparing the national inventory, but it relies on many other agencies and individuals for data, scientific information, analysis, and review. Contributors to the inventory include the Department of Energy, U.S. Department of Agriculture (U.S. Forest Service, Agricultural Research Service, National Agricultural Statistics Service), U.S. Geological Survey, Department of Transportation (including the Bureau of Transportation Statistics), Department of Commerce, Federal Aviation Administration, Department of Defense, and Colorado State University. Many private sector contractors and academic and research institutions from all sectors of the economy also contribute data and analysis to the inventory. The majority of emissions in the U.S. inventory are estimated using the highest-tier methods, and its preparation has stimulated some original research to improve basic understanding and emissions coefficients. The U.S. inventory also undergoes separate expert and public review before publication and submission to the UNFCCC.

combustion comprises the bulk (90 percent) of energy emissions.

The AFOLU sector is responsible for approximately 30 percent of global greenhouse gas emissions (Figure 2.1) but represents a much greater component of emissions in developing countries than in developed countries because of tropical deforestation and lower levels of industrial development. Agriculture contributes 15 percent of emissions in the developing world and deforestation contributes 35 percent. The figures are lower for Annex I countries, with agriculture emissions of less than 10 percent of the total and forests representing a net sink for CO_2 (note the negative sign for the Annex I forestry emission in Table 2.1). In contrast, emissions from the industrial and waste sectors in both Annex I and non-Annex I countries are an order of magnitude lower than emissions from the energy and AFOLU sectors.

A second simplifying message is that fewer than one quarter of the 185 countries in the World Resources Institute database were responsible for more than 80 percent of global emissions in 2000 (Figure 2.2). This concentration of sources has increased over the last decade with the surge in fossil-fuel emissions from a few rapidly growing, developing economies. Because developed countries and the largest emitting developing countries are likely to be the focus of mitigation efforts under a new climate treaty, their emissions will be of particular importance for monitoring.

The many activities and gases in the two most important sectors—energy and AFOLU—are cataloged in the next section. The most important gases in these sectors are CO_2, CH_4, and N_2O. Corresponding material for the industrial and waste sectors, including a discussion of CO_2, HFCs, N_2O, and CH_4, is contained in Appendix A.

TABLE 2.1 Sectoral Emissions from Annex I Countries for 2007

Sector	Fraction of Total Emissions[a]	Fraction of Sectoral Emissions by Gas					
		CO_2	CH_4	N_2O	HFCs	PFCs	SF_6
Total	1.00	0.81	0.12	0.06	0.01	0.00	0.00
Energy	0.91	0.94	0.05	0.01	0.00	0.00	0.00
Industrial processes	0.08	0.69	0.00	0.07	0.18	0.03	0.03
Solvent and other product use	0.00	0.48	0.00	0.52	0.00	0.00	0.00
Agriculture	0.08	0.00	0.47	0.53	0.00	0.00	0.00
Land use, land-use change, and forestry (LULUCF)[b]	−0.09	1.04	−0.03	−0.01	0.00	0.00	0.00
Waste	0.03	0.07	0.90	0.06	0.00	0.00	0.00
Other	0.00	0.95	0.02	0.02	0.01	0.00	0.00

[a] The disparate gases were added assuming their global warming potential with a 100-year time horizon. Each number in the table represents the sum of values from the national reports of all Annex I countries. Each value has its own confidence level, which varies with the gas and the national and sectoral source of emissions. No composite uncertainty calculations have been attempted.

[b] For Annex I countries, this sector overall is a net sink due to sequestration of CO_2. In national inventories, this is shown as negative emissions. Because emissions of CH_4 and N_2O in this sector are positive, the fraction of LULUCF that they represent is shown as a negative.

SOURCE: <http://unfccc.int/ghg_data/ghg_data_unfccc/items/4146.php>.

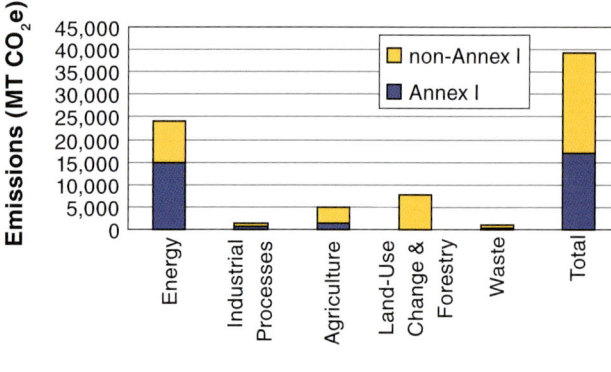

FIGURE 2.1 Greenhouse gas emissions by sector in 2000 for Annex I and non-Annex I countries; 2000 is the most recent year for which comprehensive data on the greenhouse gases are available. SOURCE: Data compiled from the Climate Analysis Indicators Tool, Version 6.0, World Resources Institute, <http://cait.wri.org/>.

SECTOR-BASED REPORTING

Energy

In most Annex I countries, CO_2 from energy use dominates anthropogenic greenhouse gas emissions. The CO_2 emissions from fossil-fuel combustion accounted for 80 percent of total greenhouse gas emissions (on a CO_2-equivalent basis) in the United States in 2006 (EPA, 2008). Other emissions from the energy sector include CO_2 from the non-energy use of fossil fuels (e.g., as petrochemicals, solvents, lubricants), CH_4 from fuel production and transport systems (e.g., coal mines, gas pipelines), and N_2O from transportation systems.

Carbon Dioxide. Most estimates of CO_2 emissions from energy systems are based on self-reporting of fuel consumption. Emissions are estimated from the amount of fuel burned, the carbon content of the fuel, and the efficiency of combustion (i.e., the fraction of fuel that is left unoxidized or incompletely oxidized at the point of combustion as, for example, carbon monoxide or ash). The fraction left unoxidized is small in modern combustion systems, and the IPCC now suggests using the default assumption that 100 percent of the carbon in a fuel is fully oxidized (IPCC, 2006). A challenge is that the amount of fuel burned is generally measured in mass or volume units and the carbon content is not generally measured. There is a good cor-

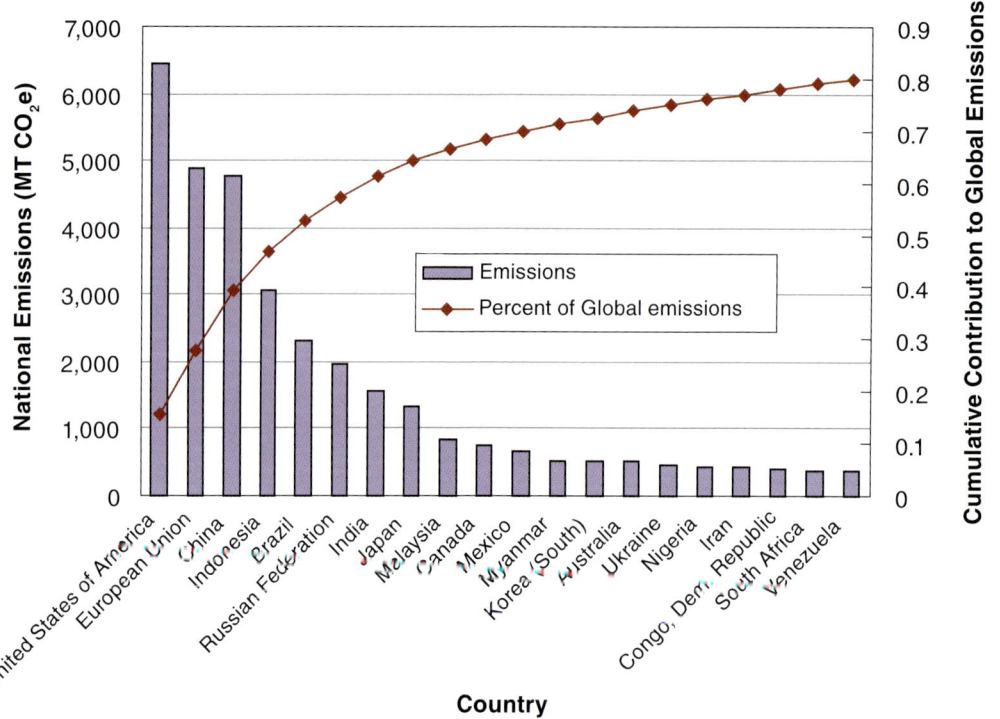

FIGURE 2.2 National greenhouse gas emissions from all IPCC sectors of the top 20 emitters in 2000. Note that the 27 countries in the European Union are treated as one. SOURCE: Data compiled from the Climate Analysis Indicators Tool, Version 6.0, World Resources Institute.

relation between the carbon content of fuels and their energy content, so conversion from mass or volume units to energy content permits the carbon content to be estimated, but with greater uncertainty (see, for example, IPCC, 2006; Marland et al., 2007; EPA, 2008, particularly Annex 2).

Direct data on fuel consumption are not always available. In some cases, CO_2 emissions can be estimated using models that represent the major fuel-consuming processes, such as the amount of CO_2 emitted per vehicle-mile of travel. At national scales, it is sometimes appropriate to estimate consumption from the amount of fuel produced and the net of imports and exports. However, fuel may not be consumed where and when it is made available. For example, fuel sold to ships, planes, or even road vehicles may be carried out of the country and burned elsewhere.[1] Data on fuel production or consumption will generally record when and where the fuel passed some point in the distribution chain. Models based on parameters such as vehicle-miles traveled or flight patterns can approximate when and where consumption takes place, but they are not widely used. Moreover, it is difficult to capture the demonstrated tendency for travelers to purchase fuel where it is cheapest (e.g., Banfi et al., 2003).

Combustion of biomass fuels is reported to the UNFCCC, but the associated CO_2 emissions are not. In theory, biomass removes CO_2 from the atmosphere when growing and releases CO_2 back to the atmosphere when burned, so a sustainably managed system should have no net CO_2 emissions. Fossil fuels used in the production, harvest, and transport of the biomass are counted in the fossil-fuel emission inventories. When sustainably produced ethanol is combined with gasoline as a fuel, the emissions counted are only those from combustion of the gasoline fraction. Similarly, if waste is used as a fuel and the waste includes both biomass and fossil-fuel-derived materials, only the CO_2 emitted from the fossil-fuel fraction is generally counted. Any net emissions of CO_2 from unsustainable use of biomass fuels should be captured as a decrease in the amount of biomass in the AFOLU sector. Thus, CO_2 emitted from the conversion of forested land to biofuels feedstock production, for example, would be included in the AFOLU inventory. This can lead to incomplete accounting of emissions if burning of biomass fuels occurs in a reporting country but the decrease in biospheric carbon stocks occurs in a nonreporting country (e.g., see Marland and Schlamadinger, 1997; Searchinger et al., 2009).

Fossil fuels are used for a variety of applications. For example, asphalt is used for roads and roofs, oils are used as lubricants and solvents, and a variety of petrochemicals are used in plastics and fibers. In the United States, more than 6 percent of the carbon in fossil fuels used in 2006 ended up in non-fuel applications (EPA, 2008), and this number can be higher in countries with large petrochemical industries. Some of the carbon used in these applications will be oxidized over time, often at slow rates (e.g., products in landfills). The IPCC provides guidance on estimating the lifetime, fate, and greenhouse gas emissions from these products. However, emissions are diffuse in both time and space, and only approximate values can be assigned to product fates and lifetimes and to changes in stocks over time.

Methane and Nitrous Oxide. Emissions of CH_4 and N_2O depend on fuel characteristics, combustion technology and maintenance, pollution-control equipment, system leakage, and prevailing current practice. Emissions occur as a result of both combustion processes and leaks from production and transport facilities such as coal mines and gas pipelines. Emissions vary widely, and emissions estimates are generally based on broad indicators and aggregate emission factors. Uncertainty is greater than for CO_2 emissions from combustion, but the sources are generally small and emissions estimates for Annex I countries have been improving with increasing interest in mitigation.

Agriculture, Forestry, and Other Land Use

The AFOLU sector is responsible for about 30 percent of global anthropogenic emissions, predominantly in the form of CO_2 emissions from land-use change (dominated by tropical deforestation) and CH_4 and N_2O emissions from farming and animal husbandry (Smith et al., 2007). Emissions vary over space and time, depending on how the land is used and on

---

[1] There is no internationally agreed-upon method for allocating fuels used in international commerce, so emissions from bunker fuels are currently reported, but not attributed.

the local climate, topography, and soil and vegetation properties. Currently, greenhouse gas emissions from land-use change are highest in tropical areas of South America, Southeast Asia, and to lesser extent, Africa (Houghton, 2003; Achard et al., 2004; DeFries et al., 2007).[2] In contrast, extratropical regions in the northern hemisphere have recently produced carbon sinks (Figure 2.3 bottom) because of net forest regrowth from earlier harvesting or encroachment on abandoned agricultural land and other processes, such as sequestration of carbon in landfills and water reservoirs and woody encroachment into pastures. These sinks are thought to absorb roughly 5-20 percent of global fossil-fuel emissions (CCSP, 2007). Methane and N_2O emissions are highest in regions with intensive, high-input agriculture (Figure 2.3 top), predominantly in Europe, North America, China, and India. Globally, agricultural emissions of CH_4 and N_2O have increased by about 1 percent per year since 1990 (Smith et al., 2007).

Estimating emissions from land-use activities is challenging due to the distributed nature of emission sources, the multitude of different processes involved, and the fact that much of the carbon is belowground in soils. Many countries simply use aggregate land-use statistics (i.e., activity data) and default emission factors (Tier 1 methods), but a number of Annex I countries use more advanced methods, employing a mixture of remote sensing, ground measurements and surveys, and process-based models. Ground surveys, where available, provide complementary information on land use and management, and ground-based measurements of carbon stock change are used to calibrate models and provide estimates of uncertainty. Models can also be calibrated with CO_2 and non-CO_2 flux measurements and related biological data. Remote sensing provides spatial information on land cover and surface characteristics (e.g., vegetation type, species, forest age; EPA, 2008) and is sometimes used to directly estimate emissions associated with land use change (e.g., Richards, 2001).

Dominant sources and sinks for the AFOLU sector include CO_2 emissions and removals from woody biomass and soils; CO_2, CH_4, and N_2O emissions from fires; CH_4 emissions from livestock and manure management and from rice cultivation; and N_2O emissions from soil management (especially nitrogen fertilization) and manure management. Other emission sources that may be locally important but are less significant globally include soil liming, organic soil cultivation, and non-rice managed wetlands or flooded lands. Long-lived harvested wood products are a potential CO_2 sink, although the average lifetime of wood products is relatively short (20 years) and UNFCCC accounting rules for them have not yet been agreed upon.

Carbon Dioxide. Net CO_2 release or uptake in managed lands is estimated from changes over time in five carbon stocks: above- and belowground biomass, dead organic matter (coarse woody debris and litter), and soil organic matter. Declines in ecosystem organic carbon stocks represent net CO_2 emissions to the atmosphere, and increases in stocks represent net CO_2 removals from the atmosphere. Carbon stock changes are calculated for six major land categories (forestland, cropland, grassland, settlements, wetlands, and other land) and for land-use change between categories (IPCC, 2006). Predominant carbon stock changes (and hence CO_2 fluxes) are associated with changes in forestland area, land use (deforestation and afforestation), and the relative carbon balance determined by growth versus harvest and natural mortality and decay. Studies of CO_2 emitted from land-use change have adopted different variables to keep track of changes in land and biomass, making it difficult to compare results directly with IPCC reporting (Ito et al., 2008).

Agricultural Methane. The largest sources of agricultural CH_4 are enteric fermentation in the digestive system of ruminants and other livestock, livestock waste, and emissions from paddy (flooded) rice. The simplest methods (Tier 1) for estimating enteric fermentation are based on emissions per animal for each major species. Countries that have additional information on livestock demographics (e.g., sex and age classes), feed quality, and intake rates can incorporate energy balance calculations to make more accurate estimates of CH_4 emissions (Tier 2). Conventional agricultural

---

[2] As a result of high tropical deforestation rates, the Conference of Parties to the UNFCCC has been considering offering economic incentives to avoid greenhouse gas emissions by changing forest management practices. See details of the COP 11 and COP 13 meetings at <http://unfccc.int/meetings/archive/items/2749.php>.

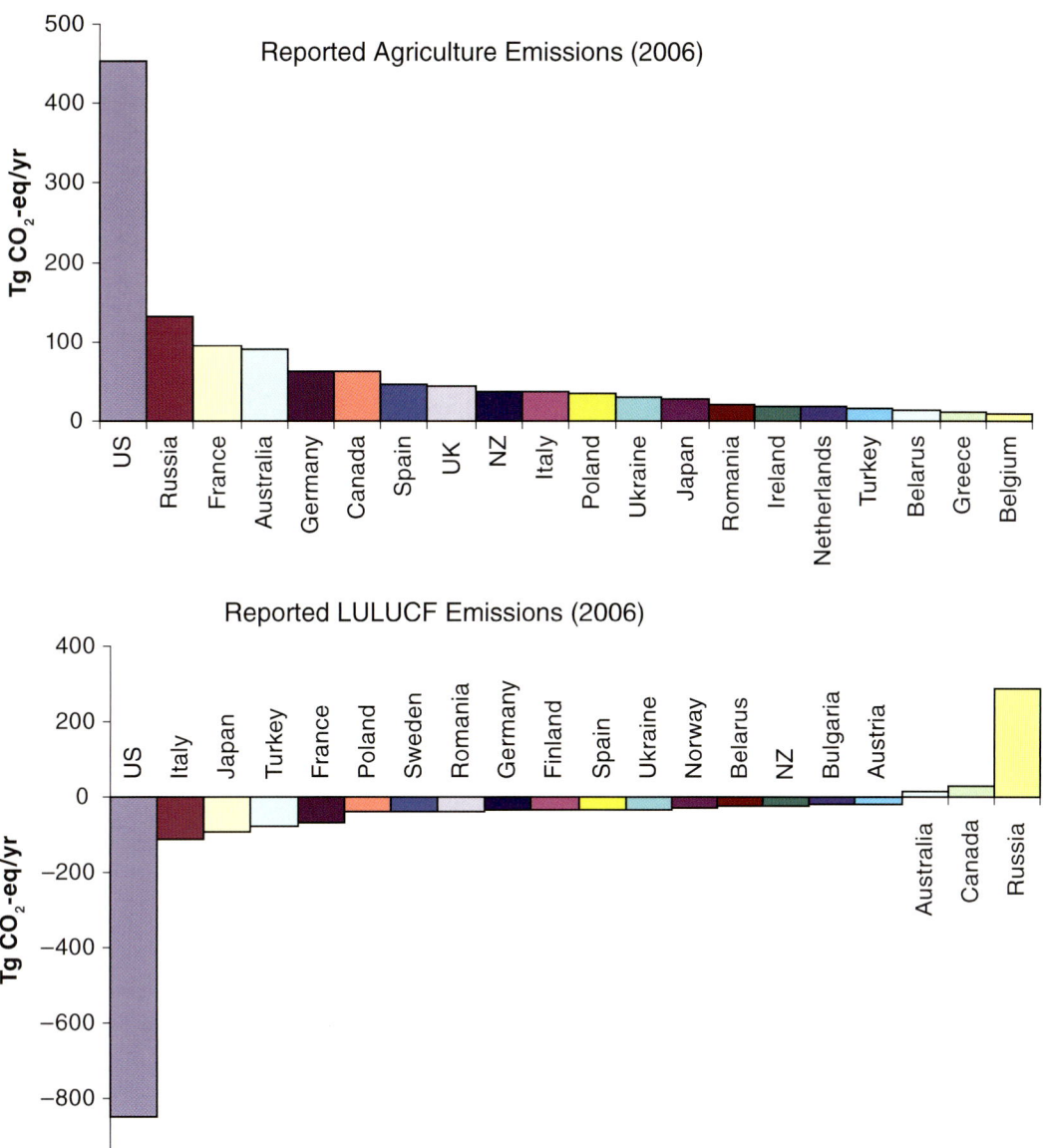

FIGURE 2.3 Annex I countries with the highest reported emissions or removals of greenhouse gases from agricultural sources (*top*) and from forestry and other land-use sources (*bottom*) in 2006. Greenhouse gases are reported as CO₂ equivalents. Negative emissions represent removals of CO₂ from the atmosphere. SOURCE: Data compiled from national greenhouse gas inventory reports; <http://unfccc.int/ghg_data/ghg_data_unfccc/>.

statistics kept in most Annex I and some non-Annex I countries provide accurate estimates of the number of animals by major species, but not always detailed information on animal nutrition. Emission factors for CH_4 from livestock waste are estimated by using manure production rates from livestock statistics, along with information on how manure is stored and managed (e.g., waste lagoons, dry lot, composted). The primary controlling factors for emissions from flooded rice are water management (length and periodicity of flooding; Li et al., 2002; Frolking et al., 2004), organic matter management (i.e., crop residues, manure; Wassmann et al., 1996; van der Gon, 1999), and cultivar type.

Agricultural Nitrous Oxide. Approximately 3-5 percent of nitrogen added through fertilization and, to a lesser extent, fossil-fuel combustion is subsequently emitted as N_2O (Crutzen et al., 2008; Galloway et al., 2008) through microbial processes in soils. Both direct emissions (i.e., those occurring at the location of nitrogen addition) and indirect emissions (i.e., those stemming from nitrogen that is volatilized or leached and redeposited elsewhere, often in nonagricultural soils, and then emitted as N_2O) are considered in AFOLU inventories. Most countries use aggregate activity data on the amount of nitrogen added to soils (from synthetic fertilizer, manure, biological nitrogen fixation from leguminous plants, and crop residues) and emission factors to estimate fluxes, although the United States and some other Annex I countries are employing more disaggregated simulation model-based approaches to reduce uncertainty. At the country level, synthetic fertilizer nitrogen production and use are well quantified from industry production and trade statistics (approximately 25 percent of global production is internationally traded) maintained by trade organizations (e.g., International Plant Nutrition Institute) and FAO. Other nitrogen addition sources such as biological nitrogen fixation and manure additions are more uncertain. Quantities of other nitrogen losses (as ammonia, nitrogen oxides, and leached nitrogen) and their contribution to indirect N_2O emissions are also highly uncertain.

Emissions from Fires. Under IPCC guidelines, CO_2 fluxes from forest fires on managed lands are incorporated in estimates of ecosystem carbon stock changes. Emissions of CH_4, N_2O, and greenhouse gas precursors (e.g., carbon monoxide, nitrogen oxides, volatile organic compounds) from incomplete combustion are inventoried separately for forests, grasslands, and croplands as a function of the area burned, pre-fire carbon stocks, and fire seasonality. The areal extent of fires can be estimated from satellite imagery (Giglio et al., 2006, 2009), although many short-duration fires and smaller (<500 m) individual fires are difficult to detect via satellite (Al-Saadi et al., 2008).

LIMITATIONS OF NATIONAL INVENTORIES FOR MONITORING

National greenhouse gas inventory reporting currently has a number of limitations for a new climate treaty. These include poor reporting by developing countries, uncertainties in reported data (which are especially high for some sources and some greenhouse gases), and lack of data to independently verify the activity data or the condition-specific emission factors used to prepare emission estimates. The problem would be most severe for a treaty that includes commitments from developing countries, many of which lack the infrastructure and capacity to collect, analyze, and manage inventory data consistently. In addition, our ability to compare inventories with data derived from other methods is limited by incomplete coverage of greenhouse gases, the time lag between the occurrence of emissions and the completion of inventories, and the lack of spatial information.

Infrequent, Incomplete, and Unreliable Reporting for Non-Annex I Countries

The establishment of rigorous reporting and review guidelines and requirements through the UNFCCC has led to the creation and steady improvement of national greenhouse gas inventories in Annex I countries over the past decade (Breidenich and Bodansky, 2009). There are still high levels of uncertainty in the estimates for biogenic sources, gases other than CO_2, and some sectors and activities with low total emissions (see below). However, uncertainties in national totals are relatively low for Annex I countries due to their well-developed statistical systems and capacity to use higher-tier methods. The fact that most of their emissions are from fossil-fuel combustion, which has low uncertainty, also reduces the overall uncertainty of these inventories. Current uncertainties in annual emissions estimates for Annex I countries are of comparable magnitude to emissions reductions commitments, and this suggests that multiyear trends could be verified if the remaining uncertainty is considered properly (Jonas and Nilsson, 2007; Swart et al., 2007).

In contrast, national inventories of many developing countries generally have greater uncertainty and are not sufficiently rigorous to enable monitoring of emissions. The low quality of national inventories in developing countries largely reflects a lack of financial, technical, and institutional capacity. Funding to prepare national inventories is provided by the Global Environment Facility,[3] but it is sporadic and insufficient to

---

[3]The Global Environment Facility is a partnership among 178 countries, international institutions, nongovernmental organizations, and private companies that provides financial assistance to help countries meet their obligations under international agreements and conventions, including the UNFCCC. See <http://www.gefweb.org/>.

enable consistent collection and processing of activity data or to maintain institutional capacity for inventory preparation over time. For example, even when international donors provide assistance in data collection and processing, the questionnaires on energy production and consumption that are distributed annually by the United Nations and IEA statistical offices are returned incomplete or not at all by many countries. Finally, land-use changes and agriculture frequently comprise a substantial source of emissions in developing countries, and reliable data in these sectors are often not available.

Uncertainty in Self-Reported Data

The uncertainty in estimates of greenhouse gas emissions from self-reported data depends on the uncertainty in both activity data and methods used to calculate the inventory. Uncertainty in activity data depends largely on a nation's commitment to data collection and processing but also on its measurement capabilities. Uncertainty in methods depends on knowledge of parameters such as the heating value and carbon content of fuels used nationally, which is used to derive emission factors, and on knowledge of biochemical processes, such as denitrification in agricultural soils. In general, estimates of greenhouse gas emissions from fossil-fuel consumption have less uncertainty than emissions from biogenic processes, such as land-use change. Uncertainties tend to be lower for emissions trends than for emissions values for a given year.

Table 2.2 shows a representative range of uncertainties for the various sources and gases as calculated and reported by Annex I countries in their national inventories. It also provides estimates of uncertainties for the emissions estimates, where available, and apportions them into contributions from uncertain activity levels and from uncertain emission factors. Estimates of uncertainty, including those in Table 2.2, have traditionally been made through expert judgment about the quality of the data used in the calculations. Indications of uncertainty can also be derived by comparing (1) estimates from different sources at a specified time, (2) estimates from a single source over time, (3) estimates made by different methods, and (4) estimates with model predictions, remotely sensed data (Marland et al., 2009; see also Chapter 3), atmospheric measurements (see Chapter 4), and proxy measures that have some correlation with emissions (e.g., trade statistics; see below).

Uncertainty in Fossil-Fuel CO_2 Emissions. UNFCCC inventories in Annex I countries produce reasonably accurate estimates for the most important emissions category: fossil-fuel CO_2. Uncertainties for Tier 1 methods are estimated to be on the order of ±5 percent, and some countries believe they can estimate emissions with even lower uncertainties using higher-tier methods (IPCC, 2006). For example, Rypdal and Winiwarter (2001) suggested that the 2σ uncertainty in annual fossil-fuel CO_2 emissions for countries with "well-developed energy statistics and inventories" may be as low as 2-4 percent. The uncertainty in estimates of fossil-fuel CO_2 emissions from the United States has been estimated to be −1 percent to 6 percent (95 percent confidence level; EPA, 2008). Analysis of the sum of Annex I reported emissions as well as some independent estimates and inverse modeling results found a 1σ uncertainty of 6 percent for fossil-fuel CO_2 (Prather et al., 2009).

In contrast, uncertainty in CO_2 emissions from developing countries is considered to be significantly higher. For countries with "less well-developed energy data systems," the uncertainty may be on the order of ±10 percent (IPCC, 2006). The 2σ uncertainty in annual estimates of fossil-fuel CO_2 emissions from China may be as high as 15-20 percent (Gregg et al., 2008).

Uncertainty in CO_2 emissions from fossil-fuel consumption arises largely from uncertainty in activity data. A comparison of annual CO_2 emission values calculated from national energy data reported to the United Nations and IEA by different analysts showed significant differences for individual countries, but no systematic bias and similar global totals (Marland et al., 1999). The mean difference was on the order of 3 percent for 19 Western European countries and 7 percent for 52 African countries, with differences ranging from 0.9 percent for the United States to 10 percent for India to 52 percent for North Korea, suggesting that the quality of activity data is a greater concern for developing countries. In a separate analysis, Gregg et al. (2008) found that revisions in energy data by the China National Bureau of Statistics, which became available

TABLE 2.2 Magnitude and Uncertainty Associated with Emission Factors, Activity Data, and Annual Emissions Estimates for the Principal Emission Source Categories in Annex I Countries

Emission Source	Anthropogenic Emissions of All Annex I Countries[a]	Range of Uncertainty for 7 Annex I Countries[b]		
		Emission Factor	Activity Data	Emission Estimate
CO_2—total anthropogenic				1
CO_2—largest sources				
Energy				
Fuel combustion	82.9%	1	1	1
Fugitive emissions from fuels	0.6%	1-5	1-2	1-5
Industrial processes				
Mineral products	2.3%	1	1	1-3
Metal production	2.0%	1	1	1-2
Chemical industry	0.6%	1-2	1	1-3
AFOLU				
Forestlands	−9.5%	2-4	1-2	2-4
Croplands	2.3%	1-4	1-2	1-4
CH_4—total anthropogenic				2-3
CH_4—largest sources				
Energy				
Fugitive emissions from fuels	4.6%	2-5	1-2	1-5
AFOLU				
Enteric fermentation	2.7%	1-3	1-3	1-3
Manure management	0.7%	1-4	1	1-5
Waste				
Solid waste disposal on land	2.1%	1-3	1	1-5
N_2O—total anthropogenic				2-5
N_2O—largest sources				
Energy				
Fuel combustion	0.7%	2-5	1	1-5
Industrial processes				
Chemical industry	0.5%	3-4	1	2-4
AFOLU				
Agricultural soils	4.0%	2-5	1-2	2-5
Manure management	0.5%	2-4	1-3	2-5
HFCs, PFCs, and SF_6—total anthropogenic				2-5
HFCs—largest source				
Consumption of halocarbons	1.1%	1-5	1	1-5
PFCs—largest source				
Aluminum production	0.2%	1-2	1-3	1-3
SF_6—largest source				
Use in electrical equipment	0.1%	1-3	1-3	2-4
Total % emissions covered	98.4%			

NOTES: 1 = <10% uncertainty; 2 = 10-25%; 3 = 25-50%; 4 = 50-100%; 5 = >100% (i.e., for the last category, we cannot be certain if the actual emissions value is a source or a sink).

[a] Reported 2006 data are from the UNFCCC's online greenhouse gas database (UNFCCC, 2008).

[b] Percentages for the largest sources are based on the 2006 greenhouse emissions data reported by seven Annex I countries—Australia, Denmark, Germany, Greece, Poland, Portugal, and the United States—selected to represent a range of institutional capabilities for compiling inventories. Uncertainty ranges were derived from the uncertainty estimates reported by the Annex I countries as part of their 2008 greenhouse gas inventory submissions for 1990 through 2006 to the UNFCCC. Each Annex I party is required to quantitatively assess and report the uncertainty of its inventories in accordance with IPCC (2000) good practice guidance. The uncertainty estimates for Denmark, Germany, Greece, and Poland were prepared for the majority of emissions source categories using the Tier 1 method, while the uncertainty estimates for Australia, Portugal, and the United States were developed using the Tier 2 method (i.e., the Monte Carlo method). The uncertainties associated with total anthropogenic emissions of each gas were available only for Denmark, Greece, Poland, Portugal, and the United States. Combined uncertainties for the emissions source categories were calculated from the uncertainty estimates given for the subcategories for each country according to the following equation (IPCC, 2000): Uncertainty(combined) = $\sqrt{([Uncertainty(1) \times Emissions(1)]^2 + [Uncertainty(2) \times Emissions(2)]^2 + \ldots + [Uncertainty(n) \times Emissions(n)]^2)} / (Emissions(1) + Emissions(2) + \ldots + Emissions(n))$.

Values are reported at the 95% confidence level. See Australian Government Department of Climate Change (2008); Danish National Environmental Research Institute (2008); EPA (2008, Annex 7); German Federal Environmental Agency (2008); Greek Ministry for the Environment (2008); Poland National Administration of the Emissions Trading Scheme (2008); Portuguese Environmental Agency (2008).

in 2006, raised the initially reported 2000 values by 23 percent. In contrast, Austria's self-reported estimates have been refined every year but have remained within a range of 2-3 percent.

Methods also make a difference in fossil-fuel CO_2 emissions estimates. For example, the IEA examined the effect of replacing the 1996 IPCC guidelines with the 2006 IPCC guidelines for fossil-fuel CO_2 and found that only the inventories that rely on default values (i.e., use Tier 1 methods) for factors such as heating values and carbon contents of fuels were affected, but inventories that used higher-tier methods were not (IEA, 2009).

Uncertainty in AFOLU Emissions. Reported uncertainties for annual CO_2 emissions from the AFOLU sector in Annex I countries range from less than 10 percent to 100 percent (Table 2.2). Independent estimates in the United States and other countries typically yield uncertainties in excess of 50 percent (Pacala et al., 2001; CCSP, 2007; Ito et al., 2008). The true uncertainty probably exceeds 100 percent in many or most non-Annex I countries.

Methodological uncertainty is more important for AFOLU emissions than for energy CO_2 emissions. Uncertainties for agricultural methane emissions are greater than 50 percent for rice cultivation due to variability in irrigation and other management practices, as well as inherent spatial and temporal variability in soil CH_4 production and consumption rates. The situation is better for CH_4 emissions associated with animal husbandry, because uncertainties in CH_4 emitted per animal are on the order of 30 percent for Tier 1 methods and 20 percent for Tier 2 methods (IPCC, 2006), and the numbers of animals are reasonably well known (Table 2.1). Methane emissions from manure are uncertain primarily because of poor documentation of manure management practices by small farms. N_2O emissions have high temporal and spatial variability due to the dynamic nature and variability of the concentration of mineral nitrogen species (e.g., NH_4^+, NO_3^-), labile organic carbon, and oxygen that largely govern N_2O emissions from soils. Consequently, uncertainties for N_2O emissions from anthropogenic nitrogen additions are high—greater than 50 percent—because of the variability in the direct and indirect emission of N_2O per unit of nitrogen added as well as uncertainty in nitrogen addition rates and management practices.

Finally, emissions from fires are highly uncertain (>100 percent) because of uncertainty in the amount of fuel actually combusted and in the trace gas emissions per unit of fuel burned (Campbell et al., 2007).

Limited Availability of Independent Data Sources for Validation and Verification

There are no truly independent sources of activity data, such as fuel use, against which data used in national greenhouse gas inventories can be compared. The United Nations, International Energy Agency, Department of Energy (DOE) Energy Information Administration, and BP Corporation create large international datasets on energy production and consumption, but all of these datasets rely primarily on the same self-reported national statistics. Data on fuel production and trade are sometimes available from corporate sources, but the most complete energy data are self-reported by countries. In many developing countries, these data are not complete or accurate or are not consistently reported.

Validation of Fossil-Fuel CO_2 Emissions. Direct measurements of CO_2 emissions are not currently used in preparing national-level inventories but could be used to validate portions of national inventories (Ackerman and Sundquist, 2008). For example, at several hundred power plants in the United States, continuous monitoring equipment measures the amount of CO_2 and other gases discharged from the smokestack (EPA, 2005). This technique is both complex and expensive and it is applied only at large point sources. Continuous monitoring of emissions is particularly useful where the fuel is heterogeneous or its delivery rate is difficult to measure. Ackerman and Sundquist (2008) compared CO_2 emissions estimates from fuel-based calculations and direct stack measurements at 828 U.S. power plants that used conventional fuels in 2004. The average absolute difference between the two sets of values was 16.6 percent, with the stack measurements giving higher values on average. However, because the stack measurements are both higher and lower than the fuel-based calculations, the two types of estimates differed by only 1.4 percent for total conterminous U.S. CO_2 emissions. Because emissions from power plants are such a large fraction of the U.S. total, resolving the differences in these two methods may be an efficient way to reduce uncertainty in national inventories (Ackerman and Sundquist, 2008).

Similarly, facility data collected to support emission trading programs could also be used to validate and improve national inventories. For example, the European Emissions Trading System created a new, independent data source, which European countries use to identify and fill gaps in national inventories and to improve country-specific emission factors (Herold, 2007). Data collected under EPA's recently adopted greenhouse gas reporting rule[4] can play a similar role in improving the U.S. inventory.

Atmospheric measurements may also be useful for verifying values or trends in CO_2 emissions (see Chapter 4). For example, problems with the energy data from China (see "Uncertainty in Fossil-Fuel CO_2 Emissions" above) had been identified earlier based on satellite measurements of trends in NO_2 column abundance (Stinton, 2001; Akimoto et al., 2006; Zhang et al., 2007; Gregg et al., 2008).

In the absence of physical measurements of CO_2 emissions, proxy data could be used to assess some trends. However, a climate treaty would change the historic relationships between emissions and proxies. For example, gross domestic product (GDP) is strongly correlated with CO_2 emissions, but this relationship changes through time, particularly after an energy price shock, and it differs by a factor of 2 or more among countries (see, for example, Raupach et al., 2007). Emissions reduction policies are designed specifically to decrease the trend in emissions, while having minimal impact on the trend in GDP. Independent data on world trade might indicate trends in CO_2 emissions in some countries and sectors. For example, 32 countries in the UN energy database burn only imported liquid fuels, and the import and export of fuel are captured in trade statistics.

Validation of AFOLU Emissions. Independent verification of changes in emission rates is difficult for all but the largest AFOLU sources because field measurements are scarce and expensive. As discussed in Chapter 3, remote sensing can be used to quantify significant changes in land use and rates of deforestation and afforestation with reasonable accuracy, and ground surveys can be used to verify carbon stock changes resulting from alterations in management. Verifying interventions to reduce enteric CH_4 emissions (e.g., genetic improvement of livestock, use of methane inhibitors) and N_2O soil emissions (e.g., altered timing, amount, and placement of fertilizer; use of nitrification inhibitors) depends on ground survey information and self-reporting. Water management (i.e., extent and periodicity of flooding of rice fields) could be monitored via remote sensing, although assessing other methane mitigation practices (e.g., improving cultivars, field additives to suppress CH_4 production) would depend on ground survey data. As outlined in Chapter 4, atmospheric sampling networks and transport modeling for agriculture-intensive regions might help to constrain emission estimates of CH_4 and N_2O from agricultural sources.

Proxy data could also be used to validate AFOLU emissions. For example, fertilizer production and import-export statistics could be used to verify overall reductions in fertilizer nitrogen use at the country scale and, together with statistics on agricultural production, to track changes in nitrogen fertilizer use efficiency as an indicator of improved management practices.

Limited Comparability with Data Derived from Other Monitoring Methods

UNFCCC inventories are difficult to compare with physical measurements because (1) they do not provide complete accounting of greenhouse gas sources and sinks, (2) geographically and temporally resolved emissions data are not reported, and (3) final emissions values are commonly not available for 2 years or more after they occur. National inventories do not include all emissions of greenhouse gases, primarily because the UNFCCC addresses only anthropogenic emissions and removals. Thus, for the AFOLU sector, only anthropogenic emissions and removals on managed lands are required to be counted.[5] Because unmanaged

---

[4] See <http://www.epa.gov/climatechange/emissions/ghgrulemaking.html>.

[5] The IPCC defines managed land as "land where human interventions and practices have been applied to perform production, ecological or social functions" (IPCC, 2006). Individual countries are permitted to define what constitutes managed lands, as long as the definition is transparent and they account for the land area consistently over time.

land is not considered, the UNFCCC inventories miss emissions and removals from natural disturbance and recovery as well as emission increases induced by climate change (e.g., increased CO_2 and CH_4 emissions in high-latitude tundra and boreal ecosystems). As a result, there is an inherent mismatch in the land-use component between UNFCCC inventories, process studies (e.g., Ito et al., 2008), and the atmospheric methods described in Chapter 4. In addition, some anthropogenic sources are not reported, either because methods to measure them have not been developed or approved (e.g., CH_4 from reservoirs) or because UNFCCC guidelines do not require their reporting (e.g., harvested wood products).

The lack of geographically resolved data is important for tracer-transport inversions, which depend on assumptions about the initial spatial and temporal pattern and magnitude of emissions. An inverse model evaluates how atmospheric observations can alter these prior estimates while maintaining the prior pattern (e.g., Gloor et al., 2000; Rayner and O'Brien, 2001; Law et al., 2003). The initially estimated pattern of emissions at the surface is critical for the analysis (e.g., Lauvaux et al., 2008). The availability of accurate geographically and temporally resolved inventory data would provide better prior estimates of emissions. If we know where and when the emissions occurred, as well as how much was emitted, the tracer-transport modeling can tell us the extent to which atmospheric measurements are compatible with the emissions estimates. Although many developed countries collect some spatially resolved greenhouse gas data to support mitigation or air quality programs, this information is not included in national inventory reports.

The delay in reporting final emissions values frustrates efforts to independently validate national inventories with real-time physical measurements, although it does not compromise the use of national inventories for treaty purposes. Much of the activity data on which emissions estimates are based is collected on survey forms from industrial firms and energy producers and consumers. It may take several years to collect and process data after the close of an accounting year. As a result, many emissions estimates are published 1 to 2 years after the emissions actually occur, and annual updates include not just another year of data but also revisions of earlier-year estimates when more accurate or more complete data become available.

NEAR-TERM CAPABILITIES FOR IMPROVING NATIONAL GREENHOUSE GAS INVENTORIES

Self-reporting has been, and is likely to continue to be, the primary means of monitoring greenhouse gas emissions and reductions under an international climate treaty. Tier 1 IPCC methods deliver national fossil-fuel CO_2 emissions estimates that are sufficiently accurate to document national multiyear trends of the dominant greenhouse gas source for the energy sector as a whole, although higher-tier methods are required to accurately estimate emissions by subsector. In contrast, Tier 1 methods do not yet produce sufficiently accurate emissions estimates from the next-largest sources: CO_2 from deforestation and CH_4 and N_2O from agriculture. However, even where more accurate Tier 2 and 3 methods exist, many developing countries lack the resources and infrastructure to use them. Improving the inventories of both developed and developing countries and enhancing self-reported data to facilitate independent verification are discussed below. Chapters 3 and 4 describe research needed to strengthen knowledge of the poorly constrained minor sectors and gases, as well as independent data on emissions that could improve confidence in reported inventories.

Building Inventory Capacity in Developing Countries

Although technical and financial assistance has helped many developing countries complete a national inventory, the challenge is to build the capacity for these countries to create complete and accurate inventories regularly and for all years. Some international efforts to improve energy statistics include a component to increase participation by developing countries. Examples include the Oslo Working Group on Energy Statistics, which is working to refine international standards and definitions (e.g., United Nations, 2008a,b,c), and IEA initiatives to harmonize energy statistics among international organizations (Karen Treanton, IEA, personal communication, July 2009).

Improving estimates of emissions from deforestation and agriculture, which are primary sources of greenhouse gases in developing countries, is challenging because many countries do not have the technical capacity to monitor these sources. International efforts to develop comprehensive global land category maps and to disseminate affordable satellite imagery could greatly improve emissions estimates in developing countries by providing country-level activity data on land use for preparation of national inventories (see Chapter 3).

In many cases, external funding and training will be required to strengthen government and research institutions in developing countries to build and retain expertise in greenhouse gas inventories. The cost of creating an ongoing capability for Tier 1 reporting in the largest emitting developing countries using existing data is relatively modest. For example, an initial investment of $450,000 per country[6] for data collection, training, and software and inventory tools could greatly improve the capacity of those countries to use higher-tier methods for inventory preparation. Additional resources on the order of $200,000 annually would be needed in countries without existing institutional capacity to maintain a permanent team of experts for inventory preparation (Mausami Desai, Environmental Protection Agency, personal communication, September 24, 2009). If only half of the 20 highest-emitting developing countries require support for institutional capacity, the cost to obtain annual estimates of their emissions would be $11 million over 5 years.

Improving Access to Data in Developed Countries

Developed countries typically have the infrastructure needed to generate reasonably accurate inventories of emissions from many significant sources and to carry out research on sources that are poorly understood (e.g., CH_4 and N_2O in the AFOLU sector). However, developed-country inventories would be improved by providing full and open access to basic data on energy systems and land use. Full and open access to international data would make it possible for any party to make estimates, compare data, or consider proxy indications. The United Nations, IEA, and FAO have established standards to ensure that the collection and reporting of publicly available data in international data compilations are uniform across countries and over time. These international compilations include some of the same data used by countries to estimate their emissions (e.g., production, trade, and consumption of energy; fuel characteristics; land use; agricultural production), but they provide supporting information that allows comparisons and correlations and they are publicly accessible. In contrast, data in some countries have limited availability (e.g., hard to find, incomplete, on paper only) or are available only at cost.

Facilitating Independent Verification of Self-Reported Emissions Data

Complete accounting of all greenhouse gas sources and sinks would enable more accurate comparisons of country totals with global and atmospheric data, facilitate regional estimates, and improve confidence in national estimates. Countries should continue to move toward more complete reporting of anthropogenic sources and sinks within the UNFCCC inventories. Initiatives needed to develop estimates of naturally occurring biogenic emissions and removals, such as the production of global land maps and research on biogeochemical cycles, are discussed in Chapter 3.

The development of spatially and temporally gridded emission datasets is also critical for improving atmospheric models, and several initiatives are currently under way. For example, the Vulcan project[7] estimates hourly emissions based on facility-level data and geographic patterns of fuel use, but so far it covers only fossil-fuel CO_2 emissions from the continental United States and only for a year. Annual gridded greenhouse gas estimates for the world are available from the Emissions Database for Global Atmospheric Research (EDGAR),[8] but they are based on disaggregated national, annual data and indicators of local emission-producing activities. These initiatives are useful begin-

---

[6]This is generally consistent with Global Environment Facility funding levels ($450,000 per country for national communications). However, inventory preparation is only one element of the national communication and only one year of data is currently required. See <www.gefweb.org/COUNCIL/GEF_C11/9.cc_brie12_review.pdf>.

[7]See <http://www.purdue.edu/eas/carbon/vulcan/index.php>.
[8]See <http://edgar.jrc.ec.europa.eu>.

nings, but they need to be significantly expanded. An international effort among developed countries to prepare and publish spatially and temporally gridded estimates of national emissions would provide critical initial data for the independent verification of national emissions, and for this effort, the disaggregated data may be more important than just the total magnitudes of emissions.

Governments also have a critical role in developing standardized methods to ensure that the gridded estimates are comparable and that they are produced at the appropriate spatial and temporal resolution to support comparison with the models. The horizontal resolution necessary is likely to be 50 to 100 km for global models, 8 km for regional models (Lauvaux et al., 2008), and 1 km for point sources. All of these methods would yield improved emissions estimates with any good initial data, even if they were not at the optimal resolution for the method. Temporal resolution, particularly the typical diurnal, weekly, and/or seasonal cycle of emissions, is also a critical piece of initial data. Because of day-night patterns in meteorology, greenhouse gases emitted at different times of the day can end up at different locations. For near-instantaneous measurements near point sources, the emissions measured are only from the last hour and depend on the duty cycle of the power plant.

RECOMMENDATIONS

The uncertainties in current estimates of emissions for the various greenhouse gases and major sources, as evaluated by the countries submitting the national reports, are summarized in Table 2.2. Uncertainties in self-reported CO_2 emission estimates are low (<10 percent) in many developed countries and could be lowered to similar levels in others by using the most accurate IPCC methods. Reducing uncertainties for N_2O and CH_4 emission estimates will also require improved activity data and emission factors, which will in some cases require research. Uncertainties in total anthropogenic N_2O are driven by the AFOLU component and are likely to remain high (10-100 percent) in the near term. For CH_4, the relative importance of energy and AFOLU sources will determine the extent to which uncertainty can be reduced. If improvements are made, uncertainties for total anthropogenic emissions of CH_4 could be as low as 10 percent in countries where emissions are dominated by energy, and as high as 50 percent in countries where emissions are dominated by AFOLU.

Extending regular UNFCCC inventories to other countries, pushing inventories to higher tiers, carrying out research to improve emission factors, and encouraging the expansion of inventories to estimate emissions at finer spatial and temporal scales should not only reduce uncertainty, but also improve opportunities for independent verification of traditional, national emissions inventories. Consequently, UNFCCC parties should strengthen self-reported national emissions inventories in the following manner:

1. Extend regular inventory reporting and review to all countries.

- Where necessary, provide sustained technical and financial support to develop and maintain institutional capacity and tools and training for preparing inventories in developing countries.
- Create a central land-use database to improve AFOLU estimates in national inventories from developing countries (see Chapter 3).

2. Continue to improve methods used by all countries.

- Support basic research on greenhouse gas emissions processes and corresponding improvements in IPCC methodologies, particularly for biogenic sources and the AFOLU sector (see Chapter 3).
- Extend top-tier (most stringent) reporting to the most important greenhouse gas sources in each country.

3. Facilitate cross-comparisons of self-reported data with data derived from other monitoring methods and develop independent data sources.

- Support the development of inventories of naturally occurring, land-based emissions and sinks for all lands, not just managed lands.
- Promote free and open access to relevant national and international statistics.

- Support IPCC, United Nations, and IEA efforts to improve energy statistics and FAO efforts to improve land-use and forestry statistics.
- Develop and implement standardized methods for preparing and publishing Annex I country inventories that are gridded at spatial and temporal resolutions appropriate for the particular greenhouse gas and source.

3

Measuring Fluxes from Land-Use Sources and Sinks

The agriculture, forestry, and other land-use (AFOLU) sector is the second-largest emitter of greenhouse gases, but the single greatest source of uncertainty. Uncertainty in carbon dioxide (CO_2), methane (CH_4), and nitrous oxide (N_2O) emitted by AFOLU activities is typically 50-100 percent or more. Much of the uncertainty associated with total AFOLU emissions is caused by uncertainty in measurements of carbon stocks associated with deforestation and rates of tropical forest cover change. For example, the standing biomass of tropical forests is uncertain by a factor of 2 (Houghton et al., 2001; Saatchi et al., 2007), and estimates of the annual flux of CO_2 released through forest clearing are uncertain by the same amount (Houghton, 2003; Achard et al., 2004; DeFries et al., 2007). For this reason, it is useful to focus on trends in AFOLU activity levels, rather than on the emissions themselves. For example, if the annually deforested area in a country decreases by a factor of 2, then, all else equal, CO_2 emissions from deforestation have also decreased by a factor of 2 (van der Werf et al., 2009a).

This chapter first describes remote sensing methods that are able to estimate the activities responsible for the majority of AFOLU emissions (e.g., deforestation) and removals by sinks. It then identifies research that the United States could undertake to reduce uncertainty in these estimates. Such methodological improvements are periodically incorporated into United Nations Framework Convention on Climate Change (UNFCCC) inventory methods developed by the Intergovernmental Panel on Climate Change (IPCC). The chapter concludes with a discussion of research that could lead to long-term improvements in remote sensing capabilities.

REMOTE SENSING

Remote sensing provides a means to survey vegetation and land surface properties over large areas. It can be used to estimate the area within each country of recently harvested forest, mature forest, pasture, and various kinds of cropland, including rice paddies, which are a dominant source of CH_4 emissions. It is also an effective means to measure fire and logging, which do not always lead immediately to detectable changes in forest cover. Time-series analysis of multiple scenes can be used to detect changes arising from deforestation, forest degradation, and afforestation. Deforestation refers to the conversion of forestland to agricultural cropland, grassland, and settlements. Degradation refers to a decrease in carbon stocks of ecosystems (e.g., through selective harvesting, draining of peatlands, or burning). Afforestation is the conversion of other land categories to forest.

In general, the transition from forest to cropland, urban areas, or pasture emits CO_2 because of the decomposition of woody debris and short-lived wood products and because the remaining material is often burned to facilitate conversion. After clearing, forests can remain an annual net source of CO_2 to the atmosphere for 5 to 20 years, with shorter time frames in warmer and wetter climates (Luyssaert et al., 2008) and in areas where the conversion process is rapid and more complete (Morton et al., 2008; van der Werf et al., 2009b). In contrast, afforestation causes a small long-term CO_2 sink because woody biomass and soil organic matter build up slowly on the site for decades to centuries. Forested land that is harvested (e.g., forest degradation) and then allowed to regenerate produces a large and short-lived CO_2 source, followed by a small and long-lived CO_2 sink.

Uncertainty of AFOLU Activities, Emissions, and Removals by Sinks

Table 3.1 summarizes the current levels of uncertainty for estimates of forest area, carbon stocks, and AFOLU emissions by remote sensing and the improvements in accuracy that could be accomplished using remote sensing and ground monitoring within several years. The most important messages in the table are the following:

1. The area of land that is deforested within a country can currently be assessed with an uncertainty of less than 50 percent over a period of 5 years (e.g., Hansen et al., 2008), but this could be improved to 10-25 percent annually by expanding efforts to analyze satellite imagery.
2. Uncertainty in remote sensing-based estimates of annual carbon fluxes from deforestation, reforestation, and forest degradation is now high (25-100 percent), but could be reduced to 10-25 percent by integrating remote sensing observations with new ground observations of forest biomass and peatland soil organic matter and biogeochemical models for spatial estimates.
3. The area of flooded soils that emit CH_4 (rice, wetlands) can be determined each year with a relatively low uncertainty (~10 percent), although uncertainty in emissions is high (50-100 percent). Ground survey data on specific management practices would provide the greatest reduction in uncertainty.
4. Uncertainty in annual N_2O emissions from managed soils is high (about 50 percent) for the best current inventory methods, and even higher (>100 percent) for developing countries. More flux measurements for different nitrogen management practices and agronomic systems, coupled with improvements in process-based models, would be the most effective action to reduce uncertainties. More information on trends in fertilizer consumption would provide insight about changing emissions when combined with information on management practices, crop production, and weather.

The next five sections describe the methods and studies behind these conclusions.

Available Satellites and Classification Methods

Although a variety of satellite and aircraft sensors have been used to map land cover and land use (e.g., see Table 3.2), Landsat is widely used and offers several advantages for regional- and country-level greenhouse gas inventory applications. First, the pixel resolution (30 m) is high enough to distinguish most plant cover characteristics, while providing a sufficiently large area of coverage per scene to allow regular global coverage to be practical. Second, the 16-day repeat cycle allows the seasonal information needed for classification purposes, such as greening and browning of grasslands, to be obtained in areas with little or no cloud cover. Third, Landsat data provide a time series of observations beginning in 1984, although the older data have some significant spatial gaps. A global map of land cover in 1990, when coverage is most complete, could serve as a baseline from which to identify subsequent changes (see Figure 3.1 for a Landsat map of the United States using data from the early to mid-1990s).

Forests can be distinguished from nonforests at accuracies of 80-95 percent using Landsat-type imagery (Table 3.1; Lu et al., 2007; GOFC-GOLD, 2008). Accuracies can be validated with in situ observations or very high resolution aircraft or satellite data at a statistically defensible subsample of locations (e.g., systematic or stratified sampling using Ikonos satellite data). The accuracy of remote sensing-derived land cover maps can be assessed in a number of ways, including comparisons with independent data collected in ground surveys. Errors are divided into those due to omission (exclusion of an area from a category in which it belongs) and commission (inclusion of an area in a category to which it does not belong), and categorized by land cover class and region. All approaches are based on statistical sampling. Land cover change also includes errors in geolocation and other factors.

TABLE 3.1 Reducing Uncertainties of Greenhouse Gas Emissions from Land Use Through Remote Sensing and Improved Monitoring Systems

Measurement	Current Annual Uncertainty[a]	Nature of Improvement	Recommendation[b]	Attainable Annual Uncertainty	Notes
Forest Area					
Northern forest area	1[c]	Systematic land cover mapping across different countries using multiple satellite data streams	A	1+	
Northern forest deforestation-afforestation	2-3[d]	Long-term continuity of ~30 m and ~1 m satellite observations; new investment to map changes at an annual scale using new change detection algorithms	B, A	1-2	Loss of LDCM would increase uncertainty for both northern and tropical deforestation rates to level 5 (greater than 100%)
Tropical forest area	1-3[e]	Same needs as for mapping northern forest area; improved access to international satellite observations	B, A	1-2	
Tropical forest deforestation-afforestation	3-4[f]	Same as above	B, A, C	1-2	
Carbon Stocks					
Northern ecosystem carbon stocks	2-4[g]	Improved access to existing country inventories; a new initiative to improve the spatial distribution of emission factors by combining P-band radar and lidar observations with inventory data	E, D, C	2	Many Annex I forest carbon inventories are not publicly available
Tropical ecosystem carbon stocks	3-4[h]	Capacity building for tropical forest and peatland inventories; the emission factor initiative described above	E, D, C	2	Forest biomass measurements have not been systematically organized; peatland areas and depths have not been accurately mapped
Ecosystem Degradation					
Logging	3-4[i]	Dedicated high-resolution (~1 m) observations for assessing logging rates in deforestation hot spots and for relating degradation patterns to Landsat observations; a wood products tracking system	B, A	2	
Fire emissions from tropical forests and peatlands	3-4[j]	Improved atmospheric emission ratios of CO/CO_2 for deforestation and peatland fires	E, F	2	An OCO rebuild with existing MOPITT and TES CO observations would enable improved estimates of peat emissions
CO_2 Emissions					
Northern ecosystem carbon fluxes	2-4[k]	Improved inventories with belowground carbon monitoring in cropland, grassland, and forests; integration of improved observations with biogeochemical models	A-E	2	Higher uncertainty levels for countries without inventories

continued

TABLE 3.1 Continued

Measurement	Current Annual Uncertainty[a]	Nature of Improvement	Recommendation[b]	Attainable Annual Uncertainty	Notes
Tropical ecosystem carbon fluxes	4–5[l]	Improved estimates of forest cover change, emission factors, measurement inventories including post-clearing land uses, and top-down constraints from column CO_2 satellite measurements; integration of improved observations with biogeochemical models	A–F	2	
CH_4 Emissions					
Rice production area	2–3[m]	Multisensor mapping (e.g., Landsat, Ikonos, SAR)	A, C	1	
Flooded rice area	3[n]	Multisensor mapping (e.g., SMAP, SSM/I, Landsat)	A	1	
Soil CH_4 emissions	3–4[o]	Ground surveys of management practices; flux measurements as a function of key management practices, including midseason drainage and fertilizer application	E	2	
Fire CH_4 emissions	4–5[p]	Continuous sampling instruments established on towers near fires and used to calibrate models and remote sensing data in relatively homogeneous areas	E, F	3	
N_2O Emissions					
Soil N_2O emissions	4–5[q]	Flux measurement network targeting different hot spots and soil, fertilizer, and manure management practices	E	3	

NOTES: CO = carbon monoxide; LDCM = Landsat Data Continuity Mission; MOPITT = Measurements of Pollution in the Troposphere; OCO = Orbiting Carbon Observatory; SAR = synthetic aperture radar; SMAP = Soil Moisture Active-Passive; SSM/I = Special Sensor Microwave Imager; TES = Tropospheric Emission Spectrometer.

[a]Uncertainty levels are 1 = <10%; 2 = 10-25%; 3 = 25-50%; 4 = 50-100%; and 5 = >100%.

[b]Recommended improvements:

    A. A new U.S. initiative to map global land cover and land-use change at an annual resolution using multiscale remote sensing data, including but not limited to the instruments listed in Table 3.2. A key part of this initiative is the design of multiscale approaches for ground truthing in rapidly changing areas, drawing on regional expertise from ground observers and high-resolution (~1 m) satellite and aircraft imagery.

    B. Successful launch and operation of the LDCM, together with immediate investment in follow-on missions for 30 m and 1 m resolution continuous data, which are crucial for monitoring logging and for improving emission factor estimates in hot-spot areas.

    C. Improved remote sensing data policies for sharing moderate- and high-resolution observations for forest cover and forest biomass monitoring (e.g., existing moderate-resolution visible and infrared reflectance measurements and radar observations from several countries cannot be downloaded for perusal or exploration by U.S. scientists).

    D. Improved data policies for sharing forest inventory observations among different countries.

    E. A new U.S. initiative to reduce uncertainties associated with emission factors related to land-use change and ecosystem degradation. This would include capacity building and investment in international scientific efforts to develop forest and peatland inventories in tropical countries, including carbon stocks associated with key post-clearing trajectories of land use. It would also involve a research program to find the most efficient ways to combine inventory observations with satellite observations (including existing P-band radar and lidar observations) to map spatial patterns of forest biomass and peatland carbon. For the fire component of forest and peatland degradation, new information of the emissions ratios of CO and CO_2 from different types of tropical burning would enable more effective use of column CO_2 and CO observations to constrain emissions. Emission factors for CH_4 and N_2O as a function of agronomic system and management practice (e.g., midseason drainage in rice agriculture) are not well characterized and represent a primary source of uncertainty in estimating regional scale emissions.

    F. Rebuilding and launching an Orbiting Carbon Observatory satellite to measure column CO_2. Fire emissions at the deforestation frontier in South America and Southeast Asia would be detectable by an OCO-like instrument. An OCO rebuild would also help to constrain estimates of emissions from peatland degradation in Southeast Asia.

[c]GOFC-GOLD (2008).

TABLE 3.1 Continued

^{d}Expert opinion based on Landsat change detection approaches taken by Masek et al. (2008) and Kennedy et al. (2007). Masek et al. (2008) report for the conterminous United States that the rate of disturbance between 1990 and 2000 was 0.9 ± 0.2% per year, primarily as a consequence of harvesting and fire.

^{e}For global humid tropical forest area, the difference between the Hansen et al. (2008) estimate for the year 2000 (1,152 Mha) and the Achard et al. (2002) estimate for the year 1997 (1,076 Mha) is relatively small (less than 10%). For tropical forest area as a whole (including forest area within 90 tropical countries), the 2005 Food and Agriculture Organization (FAO) Forest Resource Assessment estimate for the year 2000 is 1,828 Mha (Grainger, 2008). FAO assessments for this larger domain have varied considerably in different reports over the last three decades, particularly at the scale of individual countries (see Table S5 in Grainger, 2008). Estimates of year 2000 forest cover, for example, differed by more than 10% for 6 of the 12 countries with the largest amounts of tropical forest area (Grainger, 2008).

^{f}Assessments of tropical forest cover loss by Hansen et al. (2008) and the 2005 FAO Forest Resource Assessment differ by about a factor of 2 (5.4 Mha yr^{-1} vs. 12 Mha yr^{-1} for 2000-2005). Although part of this difference is associated with study domain; at a country level differences between the two approaches remain substantial, with a ~16% difference for Brazil between the two approaches and more than a factor of 2 difference for Indonesia (Hansen et al., 2008).

^{g}For some northern regions, such as the coterminous United States, forest biomass mapping approaches—which combine remote sensing and detailed inventory information—appear to be converging (see Table 11.2 of CCSP, 2007, and Table 5 of Blackard et al., 2008), implying that the uncertainty range is currently within ±25%. In other temperate and boreal regions, where inventory information is not as complete, the uncertainties are higher (Houghton et al., 2009; Goodale et al., 2002).

^{h}Regional-scale inventories in tropical regions depend on both the quality of the inventory observations and the remote sensing techniques used to extend these observations in space and time. Based on differences between extrapolation approaches, uncertainties for aboveground live biomass within the Amazon basin are within ±50% (Houghton et al., 2001; Saatchi et al., 2007). Peatlands within the Indonesian Archipelago are extensive (Page et al., 2004) but have not been systematically mapped in terms of carbon content or depth.

^{i}For the Brazilian Amazon, Asner et al. (2005) estimate wood extraction to range between 27 million and 50 million cubic meters of wood using Landsat observations to identify degraded areas. Detailed estimates using high-resolution satellite observations are not available for many other tropical countries.

^{j}Van der Werf et al. (2008) estimate emissions from Indonesia, Malaysia, and Papua New Guinea during 2000-2006 to be 128 ± 51 Tg C yr^{-1}, with a large component of the uncertainty attributable to an incomplete characterization of emission factors (CO/CO_2 ratios) for peat fires.

^{k}Expert opinion based on syntheses by Goodale et al. (2002) and CCSP (2007).

^{l}Global land-use change was estimated to be 1.2 ± 0.7 Pg C yr^{-1} for 2008 in Le Quéré et al. (2009), with uncertainties dominated by tropical forest areas. These uncertainty levels remain similar to those reported in Working Group 1 of the IPCC Fourth Assessment (Denman et al., 2007).

^{m}Expert opinion based on analysis of data on rough rice area and yield; available from the International Rice Research Institute at <www.irri.org/science/ricestat/index.asp>.

^{n}Expert opinion. Key factors influencing flooded area include water demand for other sectors and changing rice management practices (Frolking et al., 2004). Flooded areas can be detected using passive microwave satellite observations (Prigent et al., 2007).

^{o}Expert opinion based on studies by Wassmann et al. (1996), van der Gon (1999), Jain et al. (2000), and Li et al. (2002).

^{p}Campbell et al. (2007).

^{q}The committee assumed that the −40% to 70% uncertainty range for global annual agricultural N_2O emissions reported by Bouwman et al. (2002) was conservative and that uncertainties are likely to be considerably higher in many regions and individual countries where activities are not consistently measured. Key agricultural components that need to be sampled more effectively include manure generated from livestock management and different types of fertilizer application (e.g., Davidson, 2009).

Longer sampling intervals (e.g., 5 to 10 years) lead to higher rates of omissions of land cover change because of forest regrowth after harvest. The uncertainty can be reduced in many cases by using scenes collected more frequently (e.g., <1 year to 2 years) in the analysis. A global analysis using the Global Land Survey Landsat data record has been produced at a decadal interval, and efforts are under way to reduce this to a 5-year interval at least back to 1990 (Townshend et al., 2008). The reduction from 10 to 5 years will reduce omissions of land cover change, although 2-year intervals are likely to be necessary in regions where vegetation regrows quickly (e.g., wet tropical forests).

Remote Sensing of Deforestation, Forest Degradation, and Afforestation

Deforestation and Afforestation. Large-scale contiguous forest clearing is easily observed with a variety of satellite sensors that provide frequent global coverage (>250 m resolution), although significant deforestation also occurs in smaller patches <100 m^2 in size or in fine-scale linear patterns (Skole and Tucker, 1993). Landsat-type remote sensing (30 m resolution) can detect deforestation at fine scales (Huang et al., 2009). Nearly complete coverage of the globe from Landsat satellites is available for the early 1990s and 2000s. Lower resolution (1 km) Advanced Very High Resolution Radiometer (AVHRR) data on global land cover characteristics are available from 1992 to 1993.[1]

Deforestation by complete clearcut harvest or stand-replacing fire is currently being assessed with annual Landsat data in some regions (e.g., Oregon, California, Washington; Law et al., 2006; Kennedy et al., 2007). An accuracy assessment in Law et al. (2006)

---

[1] See the Global Land Cover Characteristics database at <http://edc2.usgs.gov/glcc/>.

TABLE 3.2 Current Land Remote Sensing Instruments in the Public Domain

Instrument	Measurement	Resolution and Coverage	Data Availability
Land Remote Sensing Satellite (Landsat)	Provides the longest continuous record of the Earth's continental surfaces	15-60 m, global	Landsat 7: 1999-present Landsat 5: 1984-present
Advanced Spaceborne Thermal Emission and Reflection Radiometer (ASTER)	Provides high-resolution images of the land surface, water, ice, and clouds	15-90 m, global	1999-present
Moderate Resolution Imaging Spectrometer (MODIS)	Measures biological and physical processes occurring on the surface of the Earth, in the oceans, and in the lower atmosphere	250 m-1 km, global	1999-present
Airborne Visible/Infrared Imaging Spectrometer (AVIRIS)	Measures constituents of the Earth's surface and atmosphere	5-20 m, aircraft is tasked	1998-present

showed an uncertainty of 20 percent when automated techniques were compared with air photos (Table 3.1). These methods may be suitable for countries with appropriate technical expertise and software capabilities for automation.

Forest Degradation. Remote sensing techniques can be used to identify partial biomass removals over large areas, particularly biofuels harvest (Asner et al., 2005) and selective harvest of high-grade trees in the tropics. In northern forests, degradation from logging can be

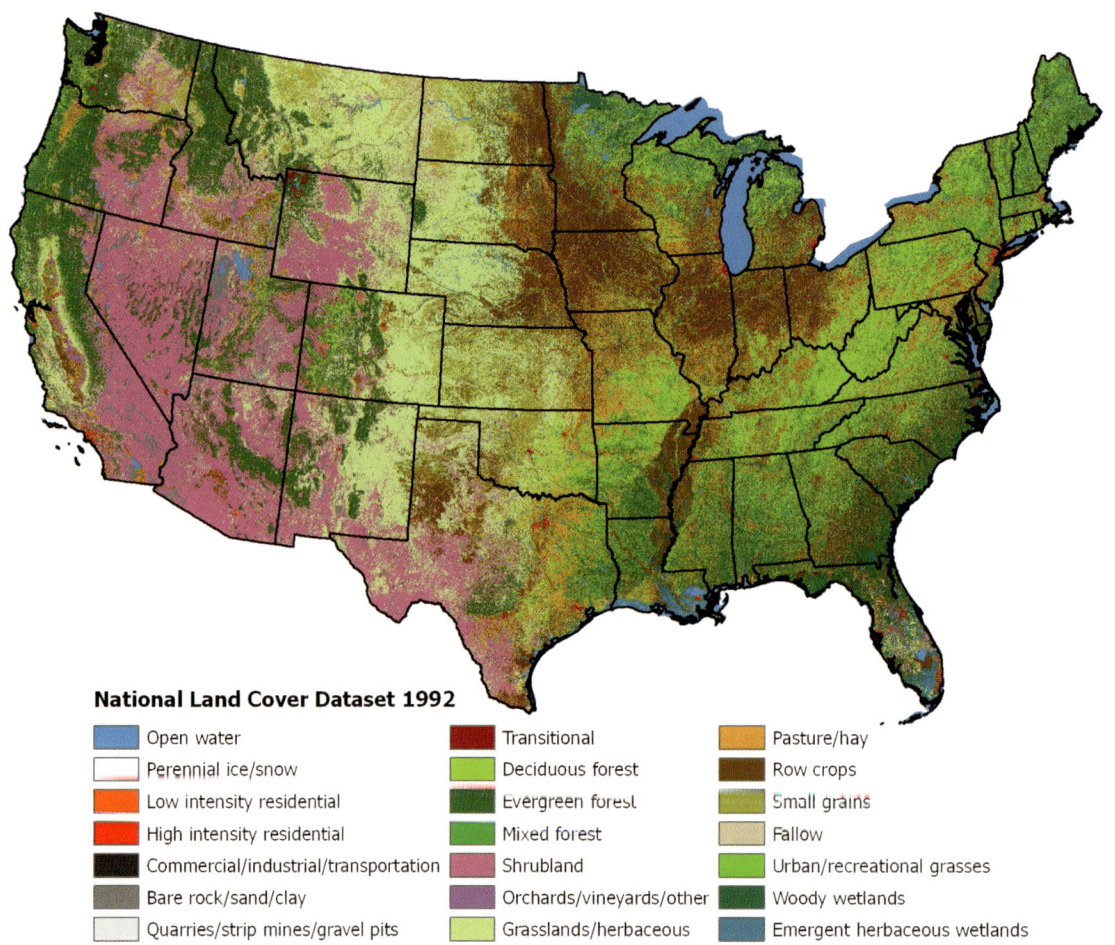

FIGURE 3.1 Major land cover features for the coterminous United States, based on early to mid-1990s Landsat Thematic Mapper satellite data. SOURCE: Vogelmann et al. (2001). Reprinted with permission from the American Society for Photogrammetry and Remote Sensing.

distinguished by inferring changes from the differences between two satellite images or, more recently, by examining as many as 25 images over a time series to identify segments that are changing over time. The former is typically done using annual to subannual Landsat data and research-level algorithms. The latter (trajectory-based image analysis), which is currently operational across the Pacific Northwest region of the United States, has advantages over traditional approaches in that it can detect forest thinning and trends such as progressive change from one land cover type to another, spreading mortality, and slow regrowth of forests over time (Kennedy et al., 2007). It can also detect a wide range of disturbance and recovery phenomena that were previously too ambiguous to label, capturing types of degradation with accuracies two to five times higher than previous change detection methods. Combining trajectory-based image analysis and high-resolution data improves the accuracy of regional estimates of terrestrial carbon fluxes and enables identification of the type of forest degradation (e.g., thinning versus mortality from insects or diseases; Kennedy et al., 2007). Such subtle disturbances may have potentially large cumulative impacts on carbon cycling at the regional scale (e.g., large-scale mortality of boreal forests from insect attack; Kurz et al., 2009).

Landsat data are also being used in time-series analysis across North America to identify forest areas subject to harvest and wildfire with a repeat interval of 2 years (Goward et al., 2008). An assessment over southeastern and northern U.S. national forests indicated overall accuracy values of 80 percent when comparing automated to human-identified disturbance mapping (Table 3.1). Most of the omissions were partial disturbances, such as thinning and storm damage, although some clearing harvests may not be detectable with temporal intervals of 2 years or more in areas of rapid forest regrowth (Huang et al., 2009). This type of approach has the potential to be applied globally.

Degradation from selective logging is more difficult to detect. In tropical forests, selective logging may leave a forest canopy that fills in within a year or that does not appear to have been thinned. The trajectory-based change detection approach has not been tested in the tropics. A new forest degradation monitoring system (Real-time Detection of Deforestation, DETER) has been developed in Brazil to detect selective logging, based on techniques of spectral mixture analysis and Normalized Difference Fraction Index (GOFC-GOLD, 2008). The system was tested in Brazil and Bolivia using 30 m Landsat data, and it failed to detect selective harvest. The system is now being tested using Satellite Pour l'Observation de la Terre (SPOT) 5 (10 m resolution) and ASTER (15 m resolution) imagery, which have spatial resolutions more appropriate for the size of individual tree canopies. Even if this approach is successful, it has three limitations: (1) it requires frequent (at least annual) mapping; (2) natural (e.g., windthrow) and human-caused degradation can be confused, possibly requiring additional ground or air photo interpretation; and (3) it requires a higher level of expertise and software for automated techniques.

At the national scale, the most effective method for detecting areas of selective harvest is to apply high spatial and temporal resolution remote sensing approaches to areas suspected of thinning, such as those determined by detection of landings along roads. Asner et al. (2005) applied an automated image analysis approach to annual Landsat data and pattern recognition techniques for detecting selective logging in the Brazilian Amazon. The analysis required initial ground-based spectroscopic characterization of surface features and tree species canopy spectra from a spaceborne hyperspectral sensor (Hyperion). The authors found an overall uncertainty of up to 14 percent in total logged area, based on seasonal Landsat data, atmospheric modeling, detection of forest canopy openings, surface debris, and bare soil exposed by forest disturbances. Alternatively, a combination of seasonal Landsat-type remote sensing and lidar or P-band radar may be required to reduce uncertainty (Treuhaft et al., 2004).

Anthropogenic fires in tropical peatlands and at the deforestation frontier contribute substantially to interannual variation in the growth rate of atmospheric CO_2 and CH_4, so fire monitoring is crucial to separate natural trends in atmospheric concentrations from the effects of mitigation. In addition, fire is used in some parts of the world to clear forest for pasture or agriculture, and fire is an important source of atmospheric CH_4 (14-88 Tg CH_4 yr^{-1}; Mikaloff Fletcher et al., 2004; van der Werf et al., 2006; Denman et al., 2007). A variety of remote sensing methods are being used to identify the location, area, and intensity of fire. The Global Observations of Forest and Land Cover

Dynamics (GOFC-GOLD) project is working to develop a global system of geostationary monitoring of active fires. Multiyear burned area products derived from moderate resolution satellite imagery are available from several sources including SPOT (Tansey et al., 2008) and the Moderate Resolution Imaging Spectrometer (MODIS; Giglio et al., 2006; Roy et al., 2008). Landsat provides more accurate mapping of fire area and severity in forests and shrublands, and it has been used for this purpose in the United States since 1984.[2] Tests using field data indicate that change detection using Landsat identifies high-severity fires with 10-30 percent uncertainty, moderate-severity fires with 40-50 percent uncertainty, and low-severity fires with 30 percent uncertainty in the western United States (Miller et al., 2009). In four fires in the western United States, low- and moderate-severity fire released 58 and 82 percent as much carbon emissions, respectively, as high-severity fire (Meigs et al., 2009), so determining the severity of all fires would reduce uncertainty in emissions estimates. National reporting is inconsistent, limiting our understanding of fire effects on forest and ecosystem degradation.

Approaches to estimating fire emissions vary widely and some result in overestimates because of faulty assumptions about the amount of fuel that is combusted (Wiedinmyer et al., 2006; Campbell et al., 2007). A common approach is to use Landsat (30 m pixels) estimates of burn area (change detection) and fire severity, and combustion completeness factors derived from field observations of live and dead biomass and surface organic matter before and after fire. Global-scale estimates use similar methods, but draw on moderate-resolution burned area observations (e.g., 500 m MODIS data) that do not differentiate fire severity. They require the use of models to estimate biomass and combustion completeness (e.g., van der Werf et al., 2006). An alternate method for global estimates is to use fire radiative power estimated at a subpixel level (e.g., using MODIS) and calibrated based on relationships between radiated energy and field estimates of combustion (Wooster et al., 2005). The results are at a coarser resolution than those yielded by the Landsat approach because active fire observations are currently available only at 1 km spatial resolution. The approach may be able to detect large homogeneous fires (27-34 percent uncertainty), but it is less effective in areas with high tree cover or heterogeneous burned areas (45 percent uncertainty; Hawbaker et al., 2008; Giglio et al., 2009). With both the regional and the global approaches described above, parameterizations of combustion completeness are now the most uncertain component of these models.

Post-fire emissions of CO_2 can be substantial and may persist for several years after fire as a consequence of decomposition of remaining organic matter, including vegetation killed but not consumed by the fire (e.g., McMillan et al., 2008). Key factors that regulate the time it takes for an ecosystem to transition from a source to a sink after fire include the severity of the fire, recruitment and growth of new plant species within burned areas, and fire-induced changes in the microclimate, which influence rates of decomposition and levels of soil moisture.

Soil Carbon

Remote sensing of soil carbon stocks is not possible because the soil immediately below the surface is opaque to the portions of the electromagnetic spectrum used to detect properties of organic matter. Further, variations in soil moisture, soil mineralogy, plant residue, and vegetation cover make even soil surface carbon estimates problematic, although low- versus high-organic-matter soils can be broadly differentiated (Sullivan et al., 2005; Yadav and Malanson, 2007). However, remote sensing provides information on aboveground vegetation (e.g., plant phenology, leaf area, photosynthetically active radiation) and residue cover. In conjunction with models, these data can be useful in estimating soil carbon stocks and stock changes and greenhouse gas emissions. Nevertheless, a much more extensive set of ground-based soil carbon measurements will be necessary to improve the soil carbon models before they are reliable enough for treaty purposes (see below).

Cropland and Pastures

Most comprehensive maps of agricultural land cover derived from satellite imagery provide information only on aggregate classes (e.g., row crops) and often at relatively coarse (e.g., 1 km) resolution (Friedl

---

[2]See <http://mtbs.gov/>.

et al., 2002). However, accurate mapping (>90 percent fidelity) of major field crop types (e.g., corn, soybean, wheat) has been demonstrated using Landsat (Daughtry et al., 2006), and higher-resolution (30-50 m) crop maps are now being generated for U.S. croplands (West et al., 2008). Recently, progress has been made in assessing crop residue coverage, which is closely correlated with tillage management, using space-borne hyperspectral instruments. For example, Daughtry et al. (2006) were able to accurately differentiate minimum (conservation) tillage fields from more intensive (reduced plus intensive) tillage practices 80 percent of the time in corn and soybean fields in central Iowa, based on comparisons with ground surveys. They used a cellulose absorption index based on reflectance in the upper shortwave infrared wavelength region from the EOS-1 Hyperion sensor.[3] Differentiation of three tillage classes (conservation, reduced, intensive) was only 60 percent accurate. Further development to correct for interference from certain types of soil minerals and to screen out pixels with more green vegetation could further improve accuracy (Serbin et al., 2009).

Methane and Nitrous Oxide

Flooded soils, including rice fields, natural wetlands, and reservoirs, are significant sources of CH_4 emissions (Denman et al., 2007). Currently, only flooded rice is included as an anthropogenic source in national inventories, although a provisional methodology for calculating CH_4 emissions from reservoirs and other artificial water impoundments is included in the IPCC guidelines. For rice methane, one of the most significant management effects is the timing and duration of flooding—a CH_4 abatement option is to reduce the period of flooding and encourage midseason drainage of paddies. Landsat (e.g., Thimsuwan et al., 2000; Biradar et al., 2008) and MODIS (e.g., Sakamoto et al., 2009) have been used to determine rice areas and phenology, and seasonal synthetic aperture radar (SAR) images (e.g., Dink-Wasser et al., 2006; Salas et al., 2007) have been used to determine the duration and timing of flooding. Studies combining remote sensing with ground-based surveys and model simulations have produced estimates of country-scale CH_4 emissions from rice in China (Li et al., 2005; Yao et al., 2006) and India (Manjunath et al., 2006). Similar techniques to map natural wetlands using remote sensing (Kaheil and Creed, 2009) and climate information could help in estimating CH_4 emissions from wetlands and other flooded soils that are not included in conventional inventories. Accuracies of 90-95 percent have been achieved in mapping and classifying different rice production systems using multiscale, multispectral satellite data (Biradar et al., 2008). However, actual CH_4 emissions have greater uncertainties (50-100 percent) at regional to national scales because they are influenced by variable water and crop residue and manure management and plant varieties, and because plant varieties must be quantified using ground survey information.

Virtually all soils emit nitrous oxide, but the main driver for increased N_2O emissions is external input of nitrogen, particularly from synthetic fertilizers but also from greater use of legume crops, manure, and deposition of the nitrate generated by combustion of fossil fuel. Emissions occur both at the site of nitrogen application (direct emissions) and in adjacent or downstream ecosystems (indirect emissions) that receive nitrogen that was lost from where it was originally applied. Remote sensing of vegetation characteristics (e.g., species, leaf area, leaf chlorophyll) that are related to plant nitrogen status can help constrain model-based estimates of N_2O (and nitrogen oxide) emissions (Martin and Asner, 2005; Vuichard et al., 2007). However, emission estimates remain highly uncertain because of the high spatial and temporal variability of fluxes and because of uncertainty in management practices related to nitrogen use and the amount and form of nitrogen in the system. Direct monitoring of nitrogen management practices via remote sensing is unfeasible, and ground surveys would be required to verify the effects of any changes in practices directed at reducing N_2O emissions (e.g., reduction in fertilizer use; change in timing, method of application, or type of fertilizer; use of nitrification inhibitors). Direct measurements of N_2O using micrometeorological techniques (e.g., Phillips et al., 2007; Fowler et al., 2009) with aircraft or tower observations, in addition to more conventional chamber methods, offer the potential for estimating emissions

---

[3] The EOS-1 Hyperion sensor was flown as a test bed for scientific applications of high spectral resolution (unlike the Landsat fixed spectral bands). It has limited spatial coverage and is near the end of its lifetime, so this satellite cannot be used for future applications.

for land areas with a high concentration of agricultural activities. Observations of NO (nitric oxide) emissions from agricultural soils following application of fertilizer and precipitation have also been detected with satellite observations (Bertram et al., 2005). Because the release of N_2O and NO is often proportional, depending on type of fertilizer (e.g., Akiyama and Tsuruta, 2003), this approach may help quantify agricultural N_2O emissions.

IMPROVING UNFCCC INVENTORIES OF LAND-USE EMISSIONS

Improvements in monitoring and verification of AFOLU emissions will depend on reducing the uncertainties of UNFCCC inventory estimates. The next three sections describe research that could leverage existing infrastructure to reduce uncertainties in AFOLU emission estimates.

Improved Carbon Inventories

Changes in forest biomass carbon stocks can be measured directly from inventories of aboveground biomass (live and dead) at two points in time, supplemented with plot data on coarse root carbon and stumps. The U.S. Forest Service Inventory and Analysis (FIA) program measures every tree on more than 100,000 plots every 5 years, and the cause of death is investigated if it died. Plots are selected using stratified random sampling of remote sensing imagery. Data on soil carbon, woody debris, and quantities such as soil nutrients, light levels, and tree health are collected on a subset of the plots. These data are generally converted to carbon using wood density data from tree cores (roughly 50 percent of stemwood biomass is carbon). The Forest Service estimates that uncertainty is 20 percent for the nation's forest carbon uptake and 10 percent for the aboveground timber volume change that is actually measured (EPA, 2008). However, estimates that include historically unmeasured pools and lands put the uncertainty at 50 percent (CCSP, 2007).

The United States does not conduct a similarly rigorous carbon inventory for nonforested ecosystems, such as croplands, pastures and natural grasslands, or shrublands. However, doing so would provide a number of advantages. An expanded inventory would provide unprecedentedly accurate estimates of CO_2 emissions from U.S. AFOLU activities (i.e., 10 percent error would be a realistic target given the experience of the FIA inventory). It would demonstrate new inventory methods that could be brought into the UNFCCC process, including better quantitative estimates of important emission factors. Finally, it would provide an accurate measure of the total CO_2 flux from non-fossil-fuel sources in the United States that could be used to develop better atmospheric methods for estimating carbon emissions (see Chapter 4).

Several European countries, Japan, Canada, and Mexico now have high-quality forest inventories, but most other nations rely on methods that are much less accurate than the FIA. Global monitoring of forest carbon stocks with ground-based inventories would provide the most accurate estimates of any method but would require an FIA-level effort in all forested countries, with spatially representative sampling, repeat visits to permanent plots, and measurements of aboveground live and dead biomass, forest floor carbon, belowground live and dead biomass, and soil carbon. Detailed methods for establishing a forest inventory are laid out in Global Terrestrial Observing System-Terrestrial Carbon Observations (GTOS-TCO) protocols (Law et al., 2008).

Global inventories of carbon fluxes from all types of ecosystems would require the implementation of similar protocols with repeated measurements of ecosystem carbon stocks over time in croplands, pastures, and nonforested natural ecosystems in all countries. This is likely to be beyond the capacity of many developing nations. For example, the cost of the FIA program in the United States is roughly $65 million per year, although FIA was designed for other purposes and the cost reflects many additional objectives. The U.S. inventory is heavily based on field data collection designed for state-level consistency in sampling intensity. An efficient design for greenhouse gas monitoring would rely more heavily on remote sensing and would have a variable sampling intensity based on ecosystem characteristics. To reduce uncertainty, measurements should include annual growth from tree cores (except some tropical species), changes in dead material (tree stems, branches, bark, stumps), and changes in soil carbon between the two measurement periods. In the United States, an expanded carbon stock measurement network for nonforested ecosystems could build on the U.S. Department of Agriculture's (USDA's) existing National Resource Inventory, which already collects an extensive set of land-use and management data, for a cost of about $5 million per year (Jeffery Goebel, USDA, personal communication, 2009). A good example of an efficient greenhouse gas inventory design is the Australian National Carbon Accounting System, which was designed explicitly for carbon accounting in the absence of an existing forest inventory and was budgeted for $35 million over 10 years.[4] The Mexican forest inventory, which is almost identical to the U.S. design but has ~17 percent of the forest area, costs ~$2 million per year (18 percent of the U.S. cost per unit forest area), largely due to significantly lower labor costs (Richard Birdsey, USDA, personal communication, August 19, 2009). Given the Australian and Mexican examples, the cost of a comprehensive national inventory of all non-fossil-fuel carbon fluxes would likely be a few (i.e., of order 10) million dollars per year in large and populous countries and significantly less in most countries, because the number of measurement sites scales with land area and the cost of labor. An assessment of monitoring costs for the Reducing Emissions from Deforestation and Forest Degradation in Developing Countries (REDD) program suggested that, depending on the policy framework and the precision needed to detect carbon stock and area changed, monitoring costs may reach $550 per square kilometer (Böttcher et al., 2009).

Eddy Flux Networks

An eddy covariance tower is a device that under most meteorological conditions measures the instantaneous exchange of CO_2 and other gases between the atmosphere and land surface for areas ranging from a hectare to a few square kilometers, depending on the height of the tower. Fluxes are computed half-hourly and summed over an entire year to provide an estimate of the annual net amount of CO_2 absorbed or released by the land ecosystem. Eddy covariance tower measurements are used to calibrate models that map annual carbon stocks and fluxes from the land. The measurements can also be used to evaluate flux estimates from ecosystem inventories. For example, records from the site at Harvard Forest show that the annual net carbon uptake over 15 years has averaged ~2.5 tons of carbon per hectare per year and has increased at an average rate of ~0.2 tons of carbon per hectare per year, which is consistent with a comprehensive carbon inventory at the same site (Barford et al., 2001). Where inventory measurements are made infrequently or turnover rates of pools are low (e.g., slow decomposition rates), it will take 5 to 10 years to compile enough data to compare with eddy covariance measurements (Curtis et al., 2002).

Eddy covariance measurements have several advantages over ecosystem inventory methods for computing the land-based net annual carbon uptake. In particular, they measure contributions to the CO_2 flux from all carbon pools, including some that inventories may miss. In addition, eddy flux sites provide information on the rapid fluctuation in carbon exchange over 24 hours and between days that is vital to constrain models of terrestrial ecosystem carbon stocks and fluxes (Urbanski et al., 2007; Medvigy et al., 2009; Wang et al., 2009) and estimates of carbon fluxes from atmospheric data (see Chapter 4).

The distribution of flux sites is determined by national scientific research programs, with a relatively large number in many developed countries, but few or none in developing countries. China and India recently started their own networks. Over the past 10 years, the number of sites in the global network has increased fivefold to 500 sites worldwide and 103 in the AmeriFlux network in the Americas. The regional networks operate independently, but protocols exist or are being developed to coordinate or standardize measurements across networks for various purposes. For example, the Integrated Carbon Observation System (ICOS) would standardize measurements at a number of European sites and combine them with other kinds of measurements to provide improved regional estimates of carbon fluxes.[5]

Eddy flux towers are too expensive (i.e., $100,000 per year) to be used to verify emissions by themselves, given the heterogeneity of ecosystems. Recall that the

---

[4] See <http://www.climatechange.gov.au/ncas/about.html> and <http://www.globalcarbonproject.org/global/presentations/2_Terrestrial/Richards.pdf>.

[5] See <http://icos-infrastructure.ipsl.jussieu.fr/>.

FIA relies on 100,000 sites to characterize stemwood dimensions of the different forest types, climate, and soils in U.S. forests. Also, the eddy covariance method is vulnerable to systematic bias errors in nonideal terrain (Finnigan et al., 2003). Nonetheless, eddy flux networks would be an essential component of an integrated system of measurements for monitoring land-use greenhouse gas fluxes in countries with sufficient capacity and funding.

Using integrated observation and modeling frameworks, annual carbon stocks and fluxes—including carbon sources and sinks from deforestation, degradation, and afforestation—can be estimated for some countries with uncertainties of ~30 percent (Luyssaert et al., 2009). Uncertainty can be reduced by using improved observations (e.g., remote sensing of disturbance history, soil carbon) and data assimilation methods in the modeling framework.

Anthropogenic Sinks at Small Scales

Under the UNFCCC, deliberately enhancing carbon uptake (e.g., by planting a forest) can be counted as an anthropogenic sink. The inventory and satellite methods described in this chapter can be used to monitor emissions and removals (sources and sinks) from forests with similar accuracy. However, verification of an emissions-trading and/or offset program would require monitoring at scales as small as a forest plantation or farm. Because ecosystem carbon uptake and release fluctuate from year to year with changes in the weather and other factors, carbon gains caused by deliberate management will be best measured against the baseline carbon flux on similar lands without the management.

Both high-resolution satellite imagery of forests and the small inventory plots used by the FIA have the spatial resolution required to assess forestry and land use at small scales. High-resolution satellites are capable of monitoring the sizes of individual trees and thus are able to monitor claimed increases in aboveground carbon stored by planted trees. Thus, conducting regular inventories of all ecosystems and satellite-based assessments of land use would also facilitate monitoring of forestry offset projects and identification of emission leakage.

Reducing Uncertainties in AFOLU Emissions of CH_4 and N_2O

The greatest impediment to reducing uncertainties in soil N_2O fluxes is the limited number of flux measurements for different climate regions, soil types, and management systems. An expanded network of flux measurement sites, using conventional chamber-based methods, at well-characterized field experiments could provide data to calibrate process-based models that integrate variable climate, soil, and management conditions. Although chamber methods are commonly used and are well suited for experimental plots comparing differences between management systems, they are subject to high spatial variability due to their small size. Combining conventional chamber-based approaches with micrometeorological techniques that are capable of estimating integrated fluxes for larger areas could further improve emission estimates (Fowler et al., 2009). Additional basic research to improve understanding of biotic and abiotic controls on N_2O fluxes and improved predictive flux models are also needed. Better quantification of indirect N_2O emissions, which are driven by the transport and subsequent emission of excess nitrogen to nonagricultural landscapes, will require flux measurements and nitrogen balance measurements at watershed scales (Deay et al., 2003). Improved survey data on soil nitrogen additions and management practices at local to national scales, which are needed to drive predictive models, can further reduce uncertainties.

Soil emissions of CH_4 are a minor component of U.S. land-use-related emissions but are a major greenhouse gas source in rice growing regions, particularly in Asia. Additional flux studies, along with better survey data on rice cropping practices (e.g., water management, residue management, manuring) are needed to reduce uncertainties in annual emissions. Although CH_4 emissions from native wetlands, reservoirs, and other flooded land are not currently required in national greenhouse gas inventories, they are highly uncertain due to the paucity of field studies conducted.

Ground-based eddy flux sites (described above) could be augmented with tunable diode laser instrumentation to measure CH_4 and N_2O fluxes (Pattey et al., 2006). The CH_4 and N_2O flux instrumentation is more expensive and difficult to manage than that required for CO_2, so deployment of such systems is currently limited, but it is likely to grow in the future (Desjardins et al., 2007; Sutton et al., 2007; Denmead, 2008).

FUTURE (>5 YEARS) OPPORTUNITIES AND THREATS

Remote Sensing Methods for Estimating Aboveground Carbon Stocks

Improvements in spatial estimates of terrestrial biomass (not including soils) are possible with planned satellite sensors. Currently, aboveground biomass is estimated using trajectory change detection maps to first identify which areas are changing and then ascribe biomass before change, biomass gained or lost, and thus biomass after the signal stabilizes. It is difficult to model biomass with Landsat alone, but Landsat time series could be combined with suitable ecosystem inventory data in a modeling framework to estimate biomass and biomass change (Samuel Goward, University of Maryland, personal communication, April 2009; Powell et al., 2010), before an orbiting vegetation lidar (light detection and ranging) sensor is flown. For example, intensive plot data can be used along with inventory tree dimension data to produce algorithms for estimating carbon stocks in vegetation and soils. The plot data are scaled spatially by developing models relating field-measured response variables to plot attributes (e.g., those related to stand age, soil fertility, climate), which are then used with satellite land cover and disturbance data and other spatial data (e.g., meteorology) to map carbon stocks (Blackard et al., 2008; Hudiburg et al., 2009). If there is a sufficient density of inventory plots, the data can be used to calibrate the remote sensing data (Landsat with lidar or P-band radar) to estimate biomass.

Other promising improvements may arise from research that combines lidar, radar, and airborne multispectral technologies (Treuhaft et al., 2004). Small (<5 m) and large (>25 m) footprint lidar have been most widely used to estimate forest carbon (e.g., Drake et al., 2002; Lefsky et al., 2002) over landscapes, but these methods also require field observations to develop the remote sensing algorithms and to assess accuracy. The multisensor approach has been applied at the landscape scale using lidar and High-fidelity Imaging Spectrometer (HiFIS) data in a tropical region with closed canopies and complex terrain (Asner et al., 2009). The method explained ~80 percent of the variation in field observations of aboveground forest biomass. The greatest source of uncertainty was in the field measurements used to develop species-specific equations. The intensive measurements, which are currently made from aircraft, are experimental and require field observations for developing the lidar equations. However, they show promise for landscape applications that could be scaled to larger areas using models or other scaling approaches (e.g., 10-year time frame).

It may be possible to calibrate the National Aeronautics and Space Administration's (NASA's) planned Deformation, Ecosystem Structure and Dynamics of Ice (DESDynI) mission and, to a lesser extent, the Ice, Cloud, and Land Elevation Satellite-II (ICESat-II) mission,[6] with aboveground biomass observations to produce estimates of aboveground carbon storage in vegetation. These could be used to monitor changes in aboveground biomass caused by major disturbances (harvest, fire, storms).

Threats to Continuity of Terrestrial Observations

Long-term trends are critical for detecting changes in vegetation caused by management or by changes in climate, atmospheric CO_2, or nitrogen deposition. However, the future continuity of observations from remote sensing and flux networks is threatened. Records from ground flux networks are now 7 to 15 years long, and they are beginning to show trends in ecosystem responses to management and climate, but there is a high risk that flux sites in the Canadian Carbon Program and CarboAfrica will be closed beyond 2010 because of budget shortfalls (Hank Margolis, Canadian Carbon Program coordinator, and Riccardo Valentini, CarboAfrica coordinator, personal communication, August 2009). In the United States, AmeriFlux sites are supported by the Department of Energy (DOE), National Oceanic and Atmospheric Administration, USDA, NASA, and National Science Foundation, with DOE supporting more than half of the current sites. Because these sites were established

---

[6]Neither mission has been scheduled for launch.

for research purposes, rather than for agency operations, there is no guarantee that they will be maintained.

Likewise, operational measurements of the land surface from space are not within the mission of any U.S. agency, jeopardizing future continuity. Key observations of land cover and land-use change have been made for more than 30 years by the Landsat series of satellites, but the two satellites in orbit are deteriorating and the Landsat Data Continuity Mission (LDCM) is not scheduled for launch until 2012. There are no firm plans to continue the series of observations after the nominal mission lifetime ends in 2018 (NRC, 2007).[7] These and additional threats to carbon cycle observations are detailed in Birdsey et al. (2009).

Access to Satellite and Inventory Observations

Improved access to satellite remote sensing and forest inventory observations is important for reducing uncertainties in carbon emissions associated with land-use change and for estimating emissions and removals from natural sources and sinks. Currently, different nations have different policies regarding access to moderate- and coarse-resolution satellite imagery. As noted above, Landsat and MODIS imagery is free, publicly available, and easily accessible. As a result, these datasets have been widely used by the international scientific community for both the development of UNFCCC AFOLU inventories and independent land-use change assessments. Both access and cost remain substantial barriers to the widespread use of satellite data from other countries, such as France (SPOT), India (India Remote Sensing Satellite), Japan (Advanced Land Observing Satellite Phased Array type L-band Synthetic Aperture Radar), and China-Brazil (Earth Resources Satellite; see Achard et al., 2007; DeFries et al., 2007; GOFC-GOLD, 2008). Moreover, data from a number of planned satellite missions (e.g., Japan's Global Change Observation Mission Second Generation Global Imager, European Space Agency's BIOMASS mission) have great potential to reduce uncertainties in carbon emissions, but are expected to carry restrictions.

An international avenue for discussing data access is the Group on Earth Observations (GEO), which is working to establish a Global Earth Observation System of Systems to improve public access to observations, datasets, tools, and expertise.[8] The group's draft carbon strategy describes using inventories, eddy covariance flux networks, atmospheric greenhouse gas observations, and ocean observations along with remote sensing observations of land cover and land-use change (Landsat, SPOT, IKONOS, MODIS, SAR) in a modeling framework to support climate agreements arising from UNFCCC conferences.[9] Open data will be required for this purpose.

RECOMMENDATIONS

The following recommendations are aimed at integrating remotely sensed land cover data with inventories and other data in a model-data framework. Landsat is emphasized because global data are freely available now and its 30-year record offers a historic baseline from which to work for treaty purposes. Lidar and the planned sensors discussed above are likely to remain in research mode for some time but may eventually be useful for supplementing Landsat-type data for validating estimates of biomass at subregional to regional scales.

1. Establish a long-term working group to produce publicly available global maps of land-use and land cover change from Landsat and high-resolution satellite imagery at least every 2 years. This will provide an independent check on the activities responsible for the majority of AFOLU emissions in self-reported UNFCCC inventories. In particular, it will enable measurement of each nation's level of deforestation, which are cumulatively responsible for approximately 12-22 percent of global anthropogenic CO_2 emissions. The maps will also provide some independent constraints on estimates of the agricultural practices (e.g., rice cultivation) responsible for a substantial fraction of CH_4 and N_2O emissions. For countries with limited capacity, the data provided under this initiative could also

---

[7] See also <http://ldcm.nasa.gov/>.

[8] As of September 2009, members of the Group on Earth Observations included 80 governments and the European Commission. See <http://www.earthobservations.org/about_geo.shtml>.

[9] See the open review version of the GEO carbon strategy at <http://www.fao.org/gtos/doc/2009-GTOS-SC/docs/8_GEO_Report.pdf>.

lay the foundation for better self-reported inventories. Global maps of land-use and land cover change could be produced by the U.S. Geological Survey, which has been disseminating land remote sensing imagery and creating Landsat data products for several decades, and/or by NASA or university scientists.

2. Increase the availability of moderate- and high-resolution satellite observations for mapping land cover change. This means that a successor to the LDCM should be added to the mission queue of NASA or another federal agency within the next year. The 30 m resolution of the Landsat instrument will have to be supplemented with 1 m imagery in a statistical sub-sampling of locations to detect and measure selective logging and to improve estimates of tree density. The high-resolution data could be obtained either by adding another instrument to the Landsat platform or by acquiring commercial or national data (with the proviso that they be made freely available). The current plan to launch a single LDCM carries with it considerable risk. If the launch fails, it would be virtually impossible for the United States to monitor land-use change using public domain information and may significantly undermine the REDD component of a future global climate treaty by limiting the capability of tropical countries to produce realistic national inventories. The implementation issues associated with maintaining a U.S. capability for collecting moderate-resolution land imaging data are discussed in FLIIWG (2007).

3. An interagency group, with broad participation from the research community, should undertake a comprehensive review of existing information and design a research program to improve and, where appropriate, implement U.S. estimates of AFOLU emissions of CO_2, N_2O, and CH_4. Key elements are likely to include continued research on the biogeochemical cycles of these gases, supported by observations from eddy covariance towers, other flux measurements for N_2O and CH_4, and ecosystem inventories of all of the major carbon pools and their trends in the United States. These observation systems will be necessary in a modeling framework (e.g., ecosystem biogeochemistry process modeling) to provide the accuracy needed for annual, spatially explicit assessments within countries.

For recommendation 3, a realistic goal is to deploy the observing systems within 5 years, which would return data needed to reduce uncertainties in AFOLU CO_2 fluxes in the United States to less than 10 percent within the following 5 years. The improved estimates could be used to provide early warning of changes in the carbon cycle that could inform the design of a climate treaty, to facilitate improvements in models of the carbon cycle, to provide data necessary to improve atmospheric methods for estimating emissions (see Chapter 4), and to demonstrate new inventory methods that could become part of the UNFCCC reporting process.

4

Emissions Estimated from Atmospheric and Oceanic Measurements

The direct way to measure emissions of greenhouse gases from a source is to collect and analyze exhaust gases as they are emitted. To estimate regional or national emissions from a remote location, one can take advantage of the incomplete mixing of the atmosphere. Because of incomplete mixing, geographical variations in greenhouse gas emissions and removals cause the abundance of a gas to be elevated downwind of a source and reduced downwind of a sink. Measurements of gas abundances around the globe can be used to estimate the locations and sizes of the gas's sources and sinks through tracer-transport inversion or inverse modeling (Box 4.1). Inverse modeling combines atmospheric and oceanic measurements with a model for atmospheric transport and mixing and models of the natural sources and sinks. Inverse modeling also works for estimates of oceanic sources and sinks, using measurements of greenhouse gases dissolved in water. Atmospheric and oceanic methods can be combined in joint atmosphere-ocean inversions (e.g., Jacobson et al., 2007a,b).

Although, in principle, tracer-transport inversion could provide independent estimates of anthropogenic emissions from individual countries for time scales of several days to a year, uncertainties using state-of-the-art methods are too high for this purpose. Four interacting factors are responsible for the large uncertainty:

1. *Small signals.* The emissions of the greenhouse gases covered in this report are small compared to the large atmospheric background. This means that anthropogenic emissions will increase the mole fraction of the gas (Box 1.3) by only a small percentage as air moves across a country (e.g., nitrous oxide [N_2O] could be increased by 1 part per billion [ppb] over a global background of 320 ppb).

2. *Incomplete understanding of atmospheric transport.* To attribute a local increase in the abundance of a greenhouse gas to emissions by an upwind country, it is necessary to have an accurate reconstruction of air flow and mixing. Errors in the atmospheric transport model will cause errors in the inferred location and magnitude of emissions.

3. *Large natural sources and sinks.* Carbon dioxide (CO_2), methane (CH_4), and N_2O all have large and uncertain natural sources and sinks that obscure the signals from anthropogenic emissions. For CO_2, the instantaneous flux into and out of the terrestrial biosphere varies with time of day and season and can be an order of magnitude larger than fossil-fuel emissions.

4. *Inadequate observing network.* Many surface stations are located too far from intense natural and anthropogenic sources to enable robust determination of global trends and seasonal cycles. Although ground stations and aircraft measurements have been deployed in Europe and North America to study CO_2 fluxes from urban and industrial regions, forests, cultivated land, and other terrestrial ecosystems, coverage of the globe is uneven and most countries are not adequately sampled.

Fortunately, technological and methodological remedies for these problems could be implemented within a few years, improving the capability for indepen-

> **BOX 4.1 Tracer-Transport Inversion**
>
> A forward model of the atmospheric and/or oceanic abundance of a trace gas is based on solution of the continuity equation:
>
> $$(X(t+\Delta) - X(t))/\Delta = \partial X(t)/\partial t = E_X(t) + P_X(t) - L_X(t) - \nabla F_X(t). \quad (1)$$
>
> The tendency in local abundance of species X at time t ($\partial X/\partial t$), written on the left as a finite difference over the time interval Δ, is equal to the local emission rate E into the volume being sampled plus in situ chemical production P minus loss rates L minus the divergence of the transport flux ∇F. These models are often designated chemistry-transport models or tracer-transport models. In a tracer-transport inversion, the measurements of $X(t)$, along with a model for the chemistry and transport ($P - L$, ∇F), are used to derive emissions (E) by subtracting $P - L$ and ∇F from both sides of equation (1).
>
> With current chemistry-transport models, this continuity equation is solved on a three-dimensional grid of more than a million points, using a meteorology (including winds, convection, diffusive mixing, clouds, and precipitation) that varies hourly. Most analyses today employ the Bayesian synthesis method (Enting, 2002; Gurney et al., 2003) that includes a priori estimates of emissions (the best estimate of the emission patterns, including uncertainties, obtained from independent data and modeling). In regions where the atmospheric measurements clearly constrain the emissions, the resulting a posteriori emissions are independent of the a priori; but in regions where the available measurements are inadequate for constraining the emissions, the method just returns what was assumed, the a priori. Thus, Bayesian methods result in more stable estimates of emissions, but sometimes add no new information.

dently verifying self-reported emissions by countries. This chapter reviews studies that used tracer-transport inversion methods to estimate CO_2 emissions and to check self-reported emissions of other greenhouse gases. The chapter also identifies four ways to reduce uncertainties associated with atmospheric monitoring of national CO_2 emissions.

INVERSE MODELING STUDIES OF GREENHOUSE GAS EMISSIONS

Tracer-transport inversion methods have been used to estimate emissions of CO_2, CH_4, N_2O, and hydrofluorocarbons (HFCs). This section summarizes the results of inverse modeling studies for estimating greenhouse gas emissions (especially CO_2) and for checking self-reported emissions (especially chlorofluorocarbons [CFCs] and HFCs) at national to global scales.

Inversions of CH_4, N_2O, and HFCs

Top-down, atmospheric inverse model derivations of greenhouse gas emissions have been applied extensively to CO_2, CH_4, N_2O, and the fluorinated gases. In general, these approaches use the patterns of variability in the trace gases to infer the geographic pattern of emissions, but they require some prior knowledge of the spatial and temporal patterns. For N_2O, the Bouwman et al. (1995) work on uncertainties in the global distribution of emissions forms the core a priori data that are tested with atmospheric observations and models (Hirsch et al., 2006; Huang et al., 2008). Thus far, these studies, as well as similar ones for HFCs (e.g., Stohl et al., 2009), have been successful at testing the assumed emissions only at the scale of broad latitudinal bands. Some studies suggest that current observations and modeling are capable of providing information on individual country emissions (e.g., Manning et al., 2003), but these claims remain untested. Many such studies assume that the atmospheric transport model represents tracer transport perfectly and do not consider the large, uncharacterized errors in the models (e.g., Patra et al., 2003; Rayner, 2004; Prather et al., 2008).

A number of studies have derived top-down emission patterns for CH_4 (e.g., Fung et al., 1991; Hein et al., 1997; Houweling et al., 1999), sometimes taking advantage of additional information contained in the relative isotopic abundances (e.g., $^{13}CH_4$ versus $^{12}CH_4$; Fletcher et al., 2004). These studies were able, for example, to verify the European Commission's Emission Database for Global Atmospheric Research (EDGAR) inventory of CH_4 emissions on a scale of North America to within an uncertainty of 20 percent, but they rely strongly on assumed patterns of emissions within the continent. Assimilation of satellite observations of CH_4 (Bergamaschi et al., 2007; Meirink et al., 2008) may be able to constrain large emissions at a sub-national level (e.g., rice paddies in India and Southeast Asia), but this has not yet been verified using a multi-model approach with realistic errors on the satellite data or by comparing predictions with regional inventories and surface observations. On the other hand, intense scientific campaigns involving regional measurements and modeling (Kort et al., 2008; Zhao et al., 2009) have succeeded in deriving CH_4 and N_2O emissions at the subnational scale within the United States, although these short, intense campaigns cannot be sustained year-round over much of the globe.

One approach to quantifying regional emissions is to use the relative increase in abundances of several gases during pollution episodes to derive their relative emissions. For example, high-resolution time series of gases have been used to identify European pollution episodes and to quantify the relative emissions of halocarbons and N_2O from Western Europe (Prather, 1985), as well as European emissions of dichloromethane, trichloroethene, and tetrachloroethene (Simmonds et al., 2006). This method has also been used to try to separate sources of CO_2 by assuming that sulfur hexafluoride (SF_6) is co-emitted with fossil-fuel combustion (Rivier et al., 2006; Turnbull et al., 2006). The approach obviates the need for accurate tracer-transport modeling and can define the relative emissions in a polluted air sample to 10 percent or better. However, it is limited to areas where distinct, single-source pollution plumes occur on top of a nearly uniform, clean air background. Surface-based, diurnal column measurements have recently been used to estimate CH_4 emissions from the Los Angeles basin (Wunch et al., 2009). Two key criteria need to be met to derive absolute emissions, for example, of fossil-fuel CO_2: (1) the absolute emissions of the reference gas must be known to equivalent or better accuracy; and (2) the two gases must be co-emitted in the same ratio over the entire region being sampled, otherwise the much larger errors in tracer-transport inversion dominate. Both criteria fail for SF_6, but for $^{14}CO_2$, the second criterion allows for the separation of fossil-fuel CO_2 from biogenic sources (see "^{14}C Measurements" below).

Overall, there are published cases for CH_4, N_2O, and the synthetic fluorinated gases in which one can distinguish source from sink and other cases in which one cannot. Thus, the committee estimates uncertainties in national emissions of these gases as 50 percent to greater than 100 percent.

CO_2 Emission Estimates

The long-term, accurate measurements of atmospheric CO_2 at Mauna Loa by Keeling (1961) and many others (Figure 1.4) laid the foundation for much of our understanding of the carbon cycle, including the link between fossil-fuel combustion and increases in CO_2 (Revelle and Suess, 1957; Pales and Keeling, 1965). Measurements of atmospheric CO_2 and other gases are currently made at in situ stations (e.g., surface sites, towers, aircraft profiling) around the world (e.g., Figure 4.1, Appendix C) and from satellites. The combined use of CO_2 and oxygen (O_2) atmospheric measurements allows the CO_2 removed from the atmosphere to be partitioned into land and oceanic carbon sinks (Keeling et al., 1993; Manning and Keeling, 2006; Denman et al., 2007). Observation of isotope analogues of CO_2 (including ^{13}C, ^{14}C, ^{18}O), CH_4 (^{13}C, ^{14}C, D), and N_2O (^{15}N, ^{18}O) as well as O_2/N_2 ratios have increased our understanding of the global sources and sinks of these greenhouse gases on scales of hemispheres or broad latitude bands. All of these measurements would significantly aid in verification or falsification of reported national emissions if the pattern of emissions being tested is provided and the measurements are made at sufficiently high spatial resolution and at suitable locations (e.g., within or close to the borders of the country whose emissions are being tested).

The mixing time of tracers in the atmosphere ranges from minutes to days between the ground and the tropopause, about two weeks around a latitude circle at midlatitudes, several months to reach through a hemisphere, about one year between the northern and southern hemispheres, and several years to mix through the stratosphere. The density of the observing network, together with the atmospheric mixing times, determines the spatial resolution with which sources and sinks can be inferred using an atmospheric model. In particular, the slow mixing between hemispheres and the preponderance of emissions in the northern hemisphere mean that the signal from anthropogenic emissions is relatively large when hemispheres are compared.

Tans et al. (1990) showed that the interhemispheric gradient in CO_2 was smaller than it should have been, given the north-south disparity in fossil-fuel emissions. This implied a "missing sink" of more than one billion tons of carbon per year in the northern hemisphere. Subsequent analyses confirmed this qualitative conclusion but have struggled to improve its spatial resolution. Continued effort has focused on the range of different atmospheric model results and what model error could

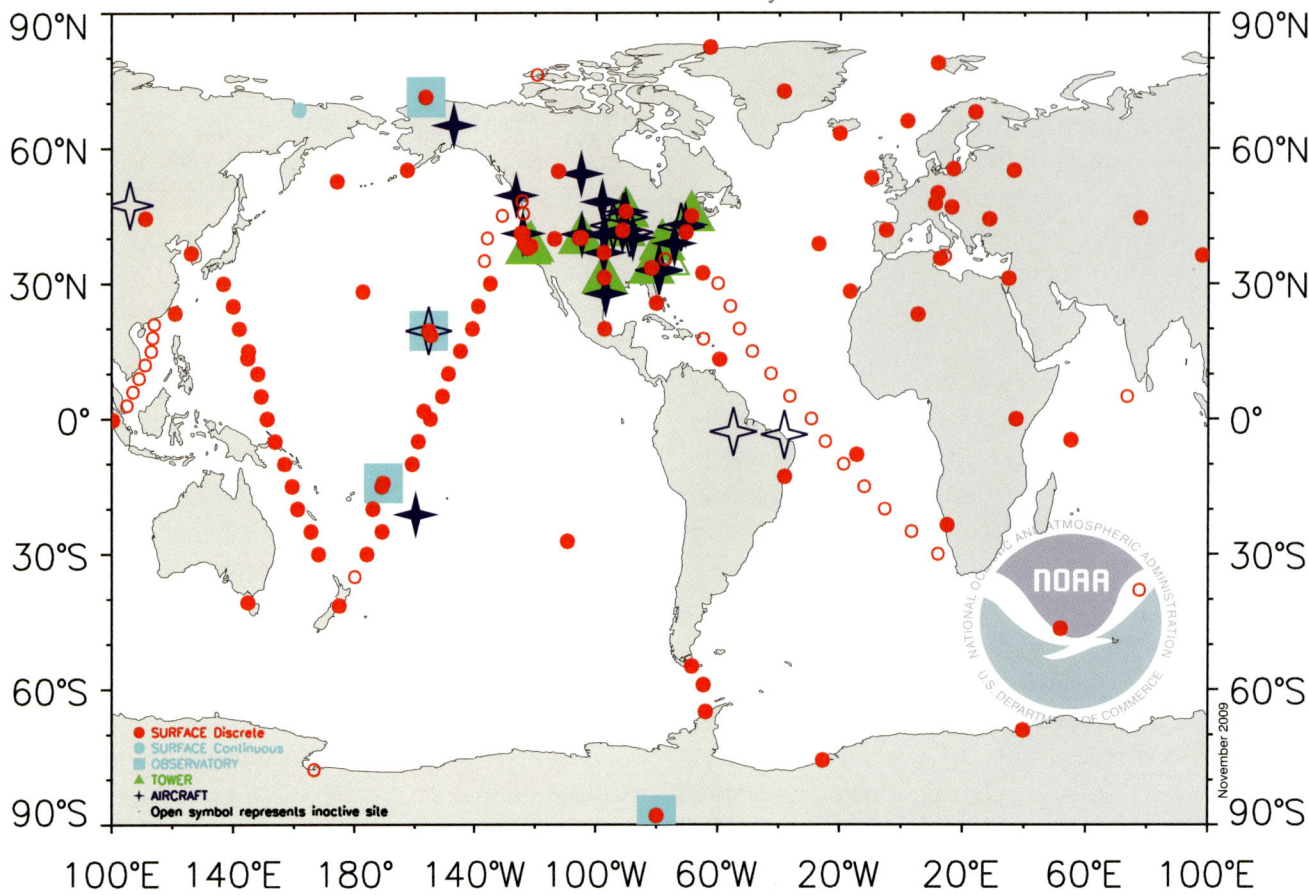

FIGURE 4.1 Map of the National Oceanic and Atmospheric Administration Earth System Research Laboratory (NOAA ESRL) global cooperative air sampling network in 2008. Blue squares are continuous measurements; red circles are weekly or daily flask samples; green triangles are continuous measurements on tall communications towers; and dark blue stars are weekly to biweekly vertical profiles by small aircraft. These sites provide a large fraction of the current set of atmospheric measurements of CO_2 and related trace gases that are used in inverse modeling. SOURCE: NOAA ESRL, <http://www.esrl.noaa.gov/gmd/ccgg>.

do to CO_2 inverse modeling (Engelen et al., 2002; Peylin et al., 2002; Gurney et al., 2003; Rayner, 2004; Baker et al., 2006a; Prather et al., 2008). The spread in north-south transport among the different models, while nonnegligible, is not large enough to invalidate interpretations of annual mean CO_2 emissions from broad latitude bands. However, the derived emissions for each continent and ocean basin within each band are highly uncertain, even in sign. The addition or loss of a single site from the observing network in Figure 4.1, for example, can in some cases shift derived emissions for large regions, such as a half-continent, from positive to negative (e.g., Rödenbeck et al., 2003; Le Quéré et al., 2007; Law et al., 2003, 2008).

Transport uncertainty causes smaller systematic errors in changes in emissions than in their absolute magnitude. For example, Denman et al. (2007) compared emissions estimates from different tracer-transport models and found that year-to-year differences in estimates for different regions within the same latitude zone were surprisingly consistent between models, despite the large between-model differences in absolute magnitude. The implication is that it is easier to quantify changes in emissions than it is to assign absolute magnitudes to those emissions.

The first row of Table 4.1 summarizes the current state of the art in CO_2 inversion estimates. It shows that although the annual atmospheric increase in CO_2 is known within 7 percent, global annual fossil-fuel emissions can be estimated from atmospheric and

TABLE 4.1 Reducing Uncertainties in CO_2 Emissions Through New Atmospheric and Ocean Observations

Method	Decadal Global Net Flux to the Atmosphere	Annual Global Net Flux to the Atmosphere	Annual Global Fossil Fuel	Annual Global IPCC LUCF	Land-ocean Sink Partition	Annual Regional Tropical Biosphere Changes	Annual Regional Mid-latitude Biosphere Changes	Annual Continental Fossil Fuel	Annual Continental IPCC LUCF	Annual National Fossil Fuel	Annual National IPCC LUCF
Current uncertainties											
State of art, including $^{13}CO_2$, O_2, oceanic, and atmospheric data	1^a (1%)	1^a (7%)	2^b (25%)	5^b	$3^{b,c,e}$	$3^{c,f}$	$3^{c,d,f}$	5	5	5	5
Potential improvements											
State of art plus $^{14}CO_2$			1-2	5^b	2^g	2^g	2^g	2^g	5	2-4	5
All of the above plus intensive aircraft observations								$++^g$		$++^g$	
All of the above plus intensive ocean observations					$++^g$	$(+)^g$	$(+)^g$	$(+)^g$			
All of the above plus other tracers of combustion (e.g., CO, NO_x, HCN)								++		++	
All of the above plus satellite CO_2 observations^b					++	++	++	++		++	

NOTES: IPCC = Intergovernmental Panel on Climate Change; LUCF = land-use change and forestry. Other than %, units are Pg of CO_2 per year for the 2000 decade.

1 = <10% uncertainty; 2 = 10-25%; 3 = 25-50%; 4 = 50-100%; 5 = >100% (i.e., cannot be certain if it is a source or sink).

Potential improvements: ++ = likely and direct; (+) = indirect in the sense that it would directly improve estimates of ocean fluxes which, in a tracer-transport inversion, would reduce errors for land masses, especially for those at similar latitude.

[a] 2σ uncertainty of annual increase, P. Tans, <www.esrl.noaa.gov/gmd/ccgg/trends/>.
[b] Denman et al. (2007).
[c] Conway et al. (1994); Ciais et al. (1995).
[d] Peters et al. (2007).
[e] Gruber et al. (2009).
[f] Tans et al. (1990); Gurney et al. (2002).
[g] Improve transport in models—expert judgment of committee.
[h] If systematic errors of satellite CO_2 retrievals can be mastered through ongoing comparisons with in situ chemical measurements, frequent resampling can lead to the small errors of annual averages that are required for flux estimates.

oceanic data only to within 25 percent. The reason for the difference is the large interannual variation in the size of sources and sinks in the terrestrial biosphere and oceans, which must be separated from the total atmospheric increase to estimate the contribution from fossil fuel. Uncertainty in the anthropogenic emissions from land-use change and forestry is greater than 100 percent because both anthropogenic and natural changes in the terrestrial biosphere have almost identical effects on atmospheric CO_2, ^{13}C, and O_2. For this reason, the inventory and remote sensing methods described in the previous chapters are needed for the a priori data on emissions from land use and forestry.

Because of transport uncertainty and the lack of measurements over much of the globe, estimates of the total net flux of CO_2 from broad bands of latitude on a seasonal time scale are uncertain by at best 25-49 percent, compared to only 7 percent for the globe (Table 4.1). The situation is far worse for estimates of continental or national emissions (>100 percent) because east-west air flow rapidly mixes emissions signals and makes inversions sensitive to model error. In addition, there remains the question of whether the amplitude of the greenhouse gas perturbations caused by national emissions is large enough to detect with in situ networks or satellites.

Appendix B presents a mass-balance analysis for the 20-largest CO_2 emitting nations. Table B.1 shows that anthropogenic emissions increase the abundance of CO_2 by, on average, a fraction of a part per million (ppm) in the whole column (ranging from 0.06 ppm for Australia to 0.76 ppm for the United States). These small signals are measurable, but their detection is confounded by the much larger and incompletely understood signals from terrestrial ecosystems. Figure 4.2 shows that, even at global scales, and when fluxes are averaged over an entire year, the apparent fraction of fossil-fuel CO_2 that remains in the atmosphere varies from year to year by as much as a factor of 2. Most of this variation is caused by the response of the terrestrial biosphere and oceans to climate anomalies (Francey et al., 1995; Keeling et al., 1995; Bacastow, 1976) and to increased fire activity during El Niño events (van der Werf et al., 2004, 2008). The magnitude of annual perturbations (sources or sinks) can be as large as one-quarter of the magnitude of global fossil-fuel emissions (e.g., Battle et al., 2000). To monitor anthropogenic emissions with tracer-transport inversion, one must be able to separate these terrestrial and ocean fluxes from the anthropogenic emissions.

Verification of CFCs, HFCs, and Other Synthetic Greenhouse Gas Emissions

The best tests of self-reported emissions using atmospheric chemistry models and observations have involved the global or hemispheric budgets of long-lived, synthetic, fluorinated gases produced solely by human activities. Rowland et al. (1982) measured CF_2Cl_2 at several remote sites to determine a global mean abundance. With knowledge of the long atmospheric lifetime (i.e., slow chemical loss), they were able to infer annual global emissions to high accuracy (e.g., 10 percent) from the annual increase in the atmosphere. The magnitude of CF_2Cl_2 emissions estimated from inverse modeling contradicted that claimed by the chemical industry, which subsequently retracted its reported emissions in favor of Rowland et al.'s derived emissions. This scenario was replayed in 2008 for the long-lived greenhouse gas NF_3, which is used in rapidly increasing quantities in the manufacture of large flat-panel displays and photovoltaic cells. Prather and Hsu (2008) reviewed the production and lifetime of NF_3, disputing an industry estimate of emissions (Robson et al., 2006) as unrealistically low and argued that this gas should be detectable and increasing in the atmosphere. Within months Weiss et al. (2008) made the measurements and confirmed that the reported NF_3 emissions were indeed too low.

Inverse modeling based on atmospheric measurements also indicates much larger emissions of HFC-134a and SF_6 than the sum of emissions in national inventories reported to the United Nations Framework Convention on Climate Change (UNFCCC). Höhne and Harnisch (2002) showed that post-1998 emissions of HFC-134a are 50 percent higher than reported by Annex I countries (see Figure 4.3), which are expected to be the source of nearly all HFC-134a emissions. Similar results have been shown for SF_6 (Geller et al., 1997; Höhne and Harnisch, 2002). Although inverse modeling shows that emissions of many HFCs and CFCs are underestimated in UNFCCC inventories (and occasionally overestimated; see discussion of halon-1301 in Clerbaux and Cunnold, 2006), it offers no insight as to the source of error. Global or hemi-

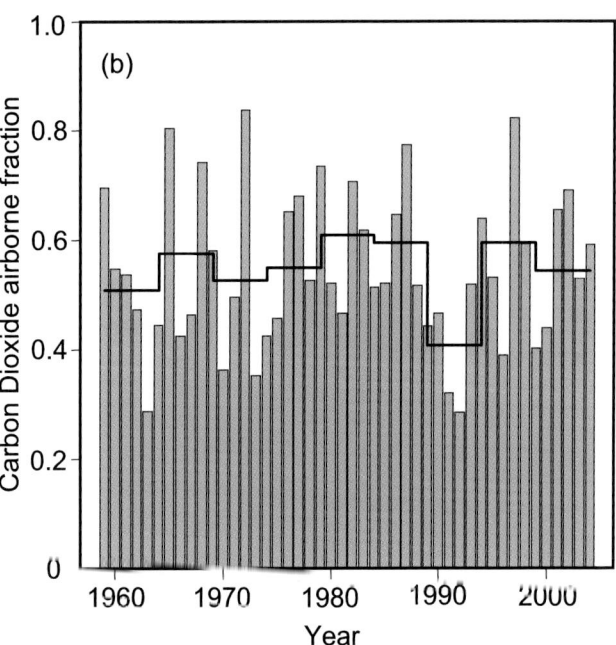

FIGURE 4.2 Interannual variation in the airborne fraction of fossil-fuel accumulation in the atmosphere. The airborne fraction is defined as the change in the mass of atmospheric CO_2 divided by the mass of fossil-fuel CO_2 emitted. The dark line shows successive 5-year averages. SOURCE: Figure 7.4b from IPCC (2007a), Cambridge University Press.

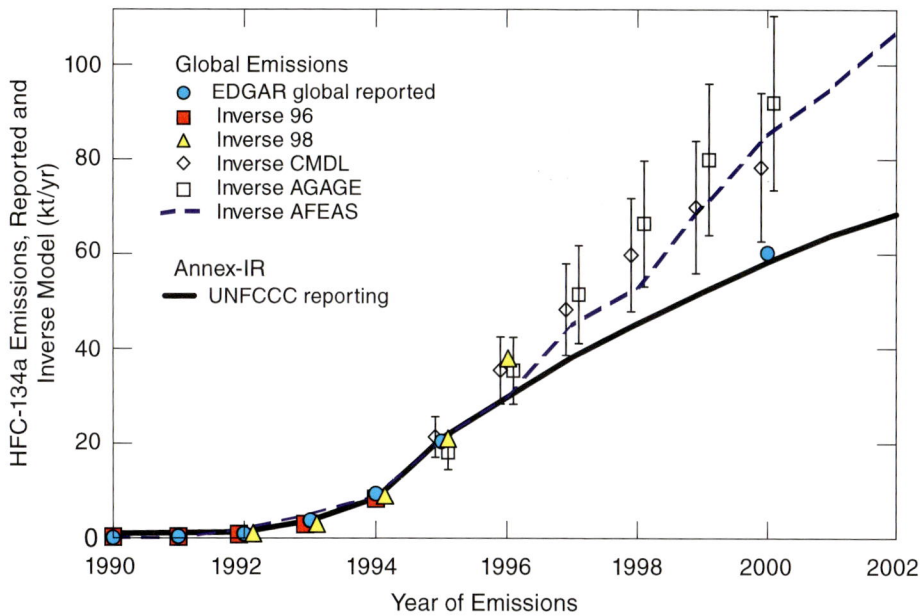

FIGURE 4.3 Global emissions of HFC-134a from several inverse atmospheric models, compared with UNFCCC reported Annex I emissions and the European Commission's Emission Database for Global Atmospheric Research (EDGAR), for 1990-2002. Most production facilities and emissions for HFC-134a are in Annex I countries. SOURCE: Courtesy of Michael Prather, University of California, Irvine. Modified from Prather et al. (2009). Copyright 2009 American Geophysical Union. Reproduced by permission of the American Geophysical Union.

spheric inverse modeling can provide a strict test of the global sum of national greenhouse gas emissions inventories on an annual basis, but it has not been able to identify the countries whose emissions are in error.

NEW APPROACHES FOR INCREASING THE ACCURACY OF NATIONAL EMISSIONS ESTIMATES

Because of the twin problems of transport error and the separation of natural from anthropogenic fluxes, the uncertainty in tracer-transport inversion estimates of anthropogenic emissions for continents and nations can be as large as 100 percent (Table 4.1). These errors currently make tracer-transport inversion impractical for monitoring national emissions. Research is improving the representation of transport and biogeochemistry in models, but at the slow pace at which new observations are becoming available, advances cannot be expected to deliver the required accuracy during the coming decade. The following approaches, which would augment current research, would shift the observing paradigm for the carbon cycle and substantially improve our capability to monitor national emissions in the near term.

Measurements of Large Emission Sources

A large fraction of fossil-fuel emissions emanates from large local sources, such as cities or power plants, and thus the effect of national mitigation measures should be evident in the "domes" of CO_2 that they produce (Idso et al., 2001; Pataki et al., 2003; Rigby et al., 2008a). For example, more than 57 percent of U.S. fossil-fuel emissions occur in areas that have a flux rate that exceeds 2 kg C m^{-2} yr^{-1}, which corresponds to ~1.7 percent of the total surface area (including power plants, cities, and other point sources; Table 4.2). Cities also

TABLE 4.2 U.S. Fossil-Fuel Emissions as a Function of Carbon Density at 0.1 Degree Grid Spacing

Carbon Density of Emissions (kg C m^{-2} yr^{-1})	Area (%)	Percentage of Total U.S. CO_2 Emissions
≤2	98.3	42.7
2-4	0.9	13.6
4-10	0.6	18.0
10-20	0.2	15.1
20-max	0.1	10.7

SOURCE: VULCAN emissions inventory; <www.purdue.edu/eas/carbon/vulcan>.

provide a broad sample of different emission sectors (Table 4.3). Statistical or systematic sampling of CO_2 emissions from large local sources would provide independent data against which to compare trends in emissions reported by the countries in which those sources are located, at least for fossil-fuel emissions. Sampling in cities, however, requires overcoming technical challenges, including finding ways to effectively construct seasonal averages in the presence of considerable spatial and daily variability and to separate biogenic from fossil-fuel sources (e.g., Pataki et al., 2003).

Working with large localized sources has two important advantages. First, their concentrated fossil-fuel emissions may be large enough to exceed the signal from local natural sources and sinks. For example, the emissions intensity of the greater Los Angeles metropolitan area (20 kg CO_2 m^{-2} yr^{-1}; see Table B.3) is ~20 times the annual net sink observed at Harvard Forest (0.9 kg CO_2 m^{-2} yr^{-1}; Barford et al., 2001), which is the difference between two much larger terms of opposite sign, photosynthesis and respiration. Of course, the sources and sinks in urban ecosystems are not as large as in Harvard Forest.

Second, large local sources increase the local CO_2 abundance in the atmosphere by a few to more than 30 ppm, depending on proximity to local sources and atmospheric mixing (see Riley et al., 2008; Mays et al., 2009; and the analysis of signals from urban areas, power plants, and leaks from a geologic sequestration site in Appendix B). Because the increased CO_2 abundances are largest over the source of emissions and disperse within a few tens of kilometers, they can usually be attributed unambiguously to their country of origin. This largely eliminates the attribution problem created by transport uncertainty in global tracer-transport inversions. For example, Figure 4.4 shows the size of the CO_2 signature of a hypothetical large power plant in the Central Valley of California, as predicted by an atmospheric model. Although the CO_2 plume at any given time varies with wind speed and direction, CO_2 mole fractions would, on average, remain elevated above a continuously emitting power plant compared with the surroundings. In this way, the CO_2 within a radius of the source could be used to infer emissions from power plants.

The CO_2 domes created by large local sources can be mapped using measurements from surface stations and aircraft in and around cities, measurements of the radiocarbon content of annual plants found in urban environments, or measurements from high-resolution satellites.

Surface Network. The variability of CO_2 and other greenhouse gases in cities is substantially greater than that measured at clean air marine boundary layer stations. Effective sampling of this variability in and

TABLE 4.3 CO_2 Emissions for Selected Cities

City	Area (km^2)[a]	Population (millions)	Emissions				
			Total (Mton CO_2 yr^{-1})	Commerce + industry (Mton CO_2 yr^{-1})	Residential (Mton CO_2 yr^{-1})	Utilities (Mton CO_2 yr^{-1})	Transportation (Mton CO_2 yr^{-1})
Los Angeles	3,700	17.5	73.2	16.8	8.1	8.8	39.9
Chicago	2,800	9.5	79.1	26.7	14.3	13.9	23.8
Houston	3,300	5.5	101.8	48.8	2.2	32.6	20.1
Indianapolis	900	2	20.1	3.7	2.2	5.5	8.8
Tokyo	1,700	29[b]	64	29.5	12.5	[c]	22
Seoul	600	13	43	18	13	[c]	12
Beijing	800	15.6	74	57	12	[c]	5
Shanghai	700	18[b]	112	92	10	[c]	10

NOTE: Mton CO_2 is million metric tons of CO_2.

[a]Area represents the contiguous area of intense activity, not administrative boundaries. Estimates of such functional boundaries were made from Google maps in "satellite" mode, which shows built up areas by color and road density.

[b]Tokyo does not include all of the agglomeration, which has 35 million inhabitants. The current population of Shanghai is 18.9 million, so emissions numbers are underestimates.

[c]CO_2 emissions are included in the commerce + industry and residential sectors.

SOURCES: Dhakal et al. (2003) for population (including migrants) and emissions in 1998 for four east Asian cities. U.S. estimates are from the VULCAN emissions inventory for 2002 (<www.purdue.edu/eas/carbon/vulcan>) and the U.S. Census.

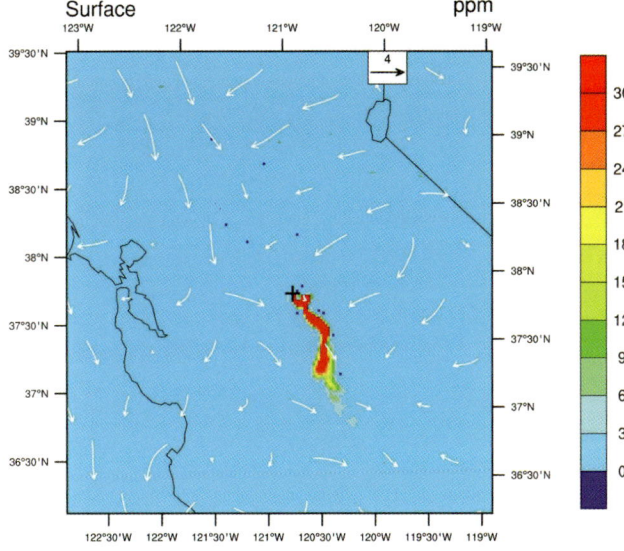

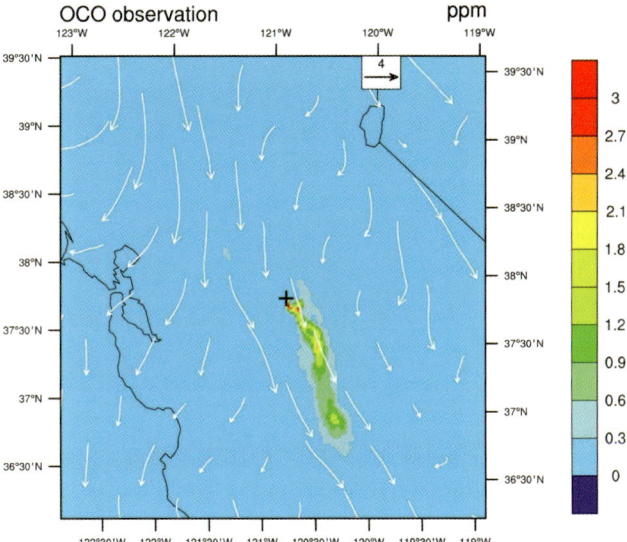

FIGURE 4.4 Instantaneous mole fraction of atmospheric CO_2 at the surface (*top*) and averaged for the column using the Orbiting Carbon Observatory (OCO) averaging kernal (*bottom*) 24 hours after emissions from a hypothetical large power plant (emitting 4.16 million tons of carbon per year) at noon in the Central Valley of California. The simulation used the Weather Research and Forecast model (WRF), with meteorology for March 4, 2008, and the model resolution was 2 km. Note the difference in scales. Wind speed and direction for the surface and for the column are shown as white arrows. Since the hypothetical point source is located in the Central Valley, high CO_2 at the surface is confined to the valley. Although there is vertical mixing, the emitted CO_2 is confined to the lower 5 km of the atmosphere and thus would not be seen by AIRS (Atmospheric Infrared Sounder), which senses the upper troposphere. SOURCE: Courtesy of Zhonghua Yang and Inez Fung, University of California, Berkeley.

around cities is needed to estimate long-term mean values and to detect multiyear changes in emissions. Care must be taken in the sampling to integrate over the daily variations in emissions and meteorological conditions. A sampling network to measure the spatial gradients in greenhouse gases created by large local sources could be established for a sample of sites. The network would require ground-based stations and aircraft sampling. Airborne technologies to support such an effort are described in Appendix D. Enhanced automated meteorological observations in urban areas would help document the dispersion of CO_2 and improve the estimation of local source strengths.

Urban Plants. Uptake of CO_2 by plants integrates the $^{14}C/^{12}C$ ratio of CO_2 in surface air over a period of several months during the growing season. The plants (by means of photosynthesis) provide long-term integrated samples during daytime periods when the planetary boundary layer is well developed and variability in CO_2 is at a minimum. Decreases in the $^{14}C/^{12}C$ ratio of plant biomass are directly proportional to the level of excess fossil-fuel CO_2 present, thus providing a seasonal measure of the dome of fossil-fuel CO_2 over major metropolitan regions (Hsueh et al., 2007; Riley et al., 2008; Pataki et al., 2010). Figure 4.5 shows the distribution of atmospheric radiocarbon anomalies in the Los Angeles basin obtained from annual grasses.

By characterizing urban to suburban gradients, it may be possible to detect relative changes in fossil-fuel emissions over time. The $^{14}CO_2$ measurements of annual plants using accelerator mass spectrometry are sufficiently accurate (2.7 permil; Riley et al., 2008) to detect a 6 percent change in fossil-fuel emissions. The actual detection limit will likely be higher because of year-to-year differences in the growing season of the sampled vegetation and variability in the synoptic-scale weather patterns. The preliminary observations shown in Figure 4.5 suggest that by sampling multiple locations repeatedly over several years, it would be feasible to detect an emissions change of ±15 percent within a large city. As with other approaches for urban air sampling, concurrent meteorological data and high-resolution tracer-transport modeling would be needed to reduce uncertainties associated with climatic variability of the atmospheric transport.

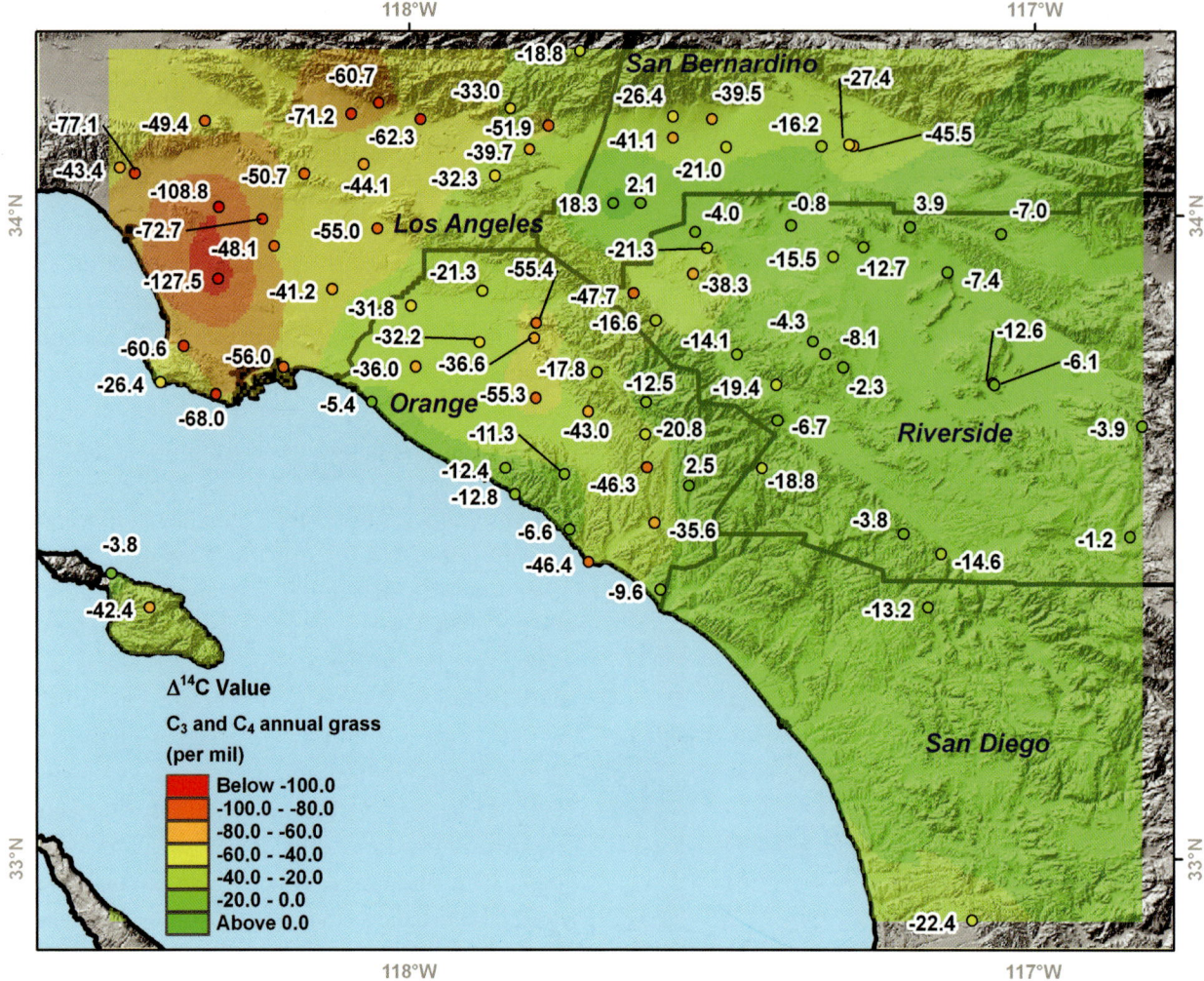

FIGURE 4.5 Radiocarbon anomalies (in permil) relative to background levels for Southern California. These anomalies were estimated relative to radiocarbon levels measured at Point Barrow, Alaska. One part per million of locally added fossil-fuel CO_2 causes a radiocarbon decrease of −2.7 permil or −0.27 percent. The largest negative anomalies were observed near downtown Los Angeles and exceeded −100 permil, corresponding to over 35 ppm of locally added fossil-fuel CO_2. Twenty sampling sites near downtown Los Angeles had a mean radiocarbon difference relative to Point Barrow, Alaska, of −58 permil, corresponding to a mean of 21 ppm of locally added fossil-fuel CO_2. SOURCE: Courtesy of Wenwen Wang and James Randerson, University of California, Irvine. Modified from Figure 19.4 from Pataki et al. (2010). Reproduced with kind permission from Springer Science and Business Media.

The approaches described above for measuring large emission sources avoid some of the problems encountered in studies that have attempted to estimate natural or agricultural fluxes using atmospheric inversion techniques. First, it is possible to directly measure the fossil-fuel emissions that create a CO_2 dome over a city or power plant. In contrast, accurate independent estimates of the actual integrated biological fluxes in a large region are not available. Second, whereas the concentration anomaly over a city or power plant has an unambiguous origin, transport errors make attribution uncertain for concentration anomalies over a natural or agricultural region.

A research program to measure the atmospheric dome of greenhouse gases over a representative sample of large local emitters would include three elements. (1) the development of new technologies and analysis approaches for detecting long-term trends in urban and industrial areas, (2) regional atmospheric modeling to help design the local network and to quantify variability from seasonally and interannually varying winds, and (3) comparison of different measurement techniques, such as flask, isotope, aircraft, sun-viewing Fourier transform spectrometer, mobile platform, and satellite. Long-term monitoring could begin in approximately 10 different U.S. cities with different sizes, population densities, and different mixtures of emissions sources using currently available sampling technologies. Simultaneous creation of detailed bottom-up inventories of emissions for these same representative areas would facilitate transport modeling and enable the efficacy of different atmospheric approaches for trend detection to be assessed. By analogy with research initiatives of similar scope, this initiative would require a research budget of about $15 million to $20 million per year (e.g., $6 million to $10 million for extending National Oceanic and Atmospheric Administration [NOAA] flask and continuous monitoring to 10 cities and the remainder for technology development, modeling, and analysis) in the United States. Ideally, similar programs would be developed in other countries.

Satellites. By providing repeated global coverage from a single instrument, satellites would complement a global network of ground stations and aircraft sampling to monitor CO_2 emitted from selected cities and power plants. They would also largely overcome the difficulties of national sovereignty and international cooperation in CO_2 monitoring and the verification challenges associated with self-reporting. As shown in Table C.1 (Appendix C), Japan's Greenhouse gases Observing Satellite (GOSAT) is the best available satellite for spaceborne measurement of CO_2 anomalies from emissions. It has lower uncertainty and higher spatial resolution than the Scanning Imaging Absorption Spectrometer for Atmospheric Chartography (SCIAMACHY), the Atmospheric Infrared Sounder (AIRS), or the Infrared Atmospheric Sounding Interferometer (IASI), and it senses near the surface where emission signals are largest, unlike AIRS and IASI. However, the CO_2 signal produced by the emissions of a large power plant is typically smaller than what can be measured with GOSAT. In contrast, the Orbiting Carbon Observatory (OCO), which failed on launch in February 2009, would have had the high spatial resolution required to monitor instantaneous CO_2 emissions from such local sources.

For example, assume that a 500 MW pulverized coal power plant emits ~0.13 t s^{-1} of CO_2 (e.g., 4 Mt CO_2 yr^{-1}) and that the wind speed is 3 m s^{-1}. These conditions would produce a perturbation of approximately 0.5 percent (~1.7 ppm) in the average column abundance of CO_2 within an OCO sample, which is consistent with the instrument's design uncertainty of 1-2 ppm and significantly larger than the ground-tested value of 1 ppm. In contrast, because a GOSAT sample covers a larger area than an OCO sample, the CO_2 perturbation within a GOSAT sample would be approximately 0.1 percent (~0.4 ppm). This is an order of magnitude smaller than GOSAT's estimation error of 4 ppm. Note that in target mode, OCO could take up to 7,000 shots at an individual site, under different viewing angles, and could potentially have an uncertainty of 0.1 ppm if systematic biases were characterized and removed.

No other satellite has OCO's critical combination of high precision, small footprint, readiness, density of cloud-free measurements, and ability to sense CO_2 near the Earth's surface (Table C.1, Appendix C). Its 1-2 ppm accuracy and 1.29 × 2.25 km sampling area would have been well matched to the size of a power plant. Yet, the OCO mission as planned would have had limitations for monitoring CO_2 emissions from large sources because it would have sampled only 7-12 percent of the land surface (Miller et al., 2007) with a revisit period of 16 days and a nominal lifetime of only 2 years (Table C.1, Appendix C). However, many metropolitan areas are large enough to be sampled by the planned orbit, and OCO would still have provided a sample of a few percentage of the power plants.

Monitoring urban and power plant emissions from space is challenging and has not been demonstrated. A replacement OCO could demonstrate these capabilities. Nevertheless, it would be valuable to explore changes in the orbit and other parameters so that a greater fraction of large sources is sampled. For example, consider a precessing orbit covering ~100 percent of the surface but with only two measurements per year of each location. With 100-500 large local sources in high-emitting countries, it might be possible to obtain a statistical sample of hundreds of measurements of plumes of CO_2 being emitted by the large sources in each of these countries. The trade-offs in optimizing monitoring capabilities while meeting scientific objectives (e.g., observing with high solar zenith angle) would have to be examined by a technical advisory group. The group would also have to develop a strategy for integrating the snapshot view of CO_2 anomalies from the satellite measurements into annual emissions.

Because of its 2-year mission life, OCO would not have been able to track emission trends. However, it would have provided the first few years of measurements (a baseline) necessary to verify a decadal trend for the large local sources within its footprint and served as a pathfinder for successor satellites designed specifically to support a climate treaty. Moreover, the technology used in the OCO instrument has no consumables or inherently short-lived components and hence could be flown on an extended, decadal mission. A replacement mission is expected to cost about the same as the original, $278 million.[1] Alternate proposals to measure the near-surface CO_2 abundance from space with multifrequency laser (e.g., the Active Sensing of CO_2 Emissions over Nights, Days, and Seasons [ASCENDS] mission)[2] are not yet technologically ready for long-term operations.

Deployment of a CO_2-sensing satellite with the capability of OCO, together with a program of surface and aircraft sampling in and around large local sources would address all four main problems associated with tracer-transport inversion. It would enable the network to support emissions verification by observing CO_2 directly over high-emitting sites, where the signal from anthropogenic emissions is locally large enough to separate from confounding natural fluxes and where the transport problem is far simpler than at regional and continental scales because emissions are in most cases still confined to the boundary layer. Ongoing sampling by aircraft or from balloons will be essential to investigate potential systematic biases of column abundances inferred from any satellite and to describe the vertical distribution necessary to infer the location of emissions.

A corresponding capability for other greenhouse gases does not exist, except for the ozone precursor NO_x (nitrogen oxide). Satellite remote sensing of NO_x emissions is readily able to detect decadal emission trends over urban and industrial regions (Martin et al., 2006; Stavrakou et al., 2008; Kaynak et al., 2009).

---

[1]See <http://oco.jpl.nasa.gov/news/index.cfm?FuseAction=ShowNews&NewsID=37>.

[2]See <http://decadal.gsfc.nasa.gov/ascends.html>.

^{14}C Measurements

Systematic radiocarbon (^{14}C) measurements would largely eliminate the confounding effects of natural fluxes and thus greatly improve the accuracy of fossil-fuel emissions estimates. Carbon-14 is produced by cosmic rays in the higher layers of the atmosphere as well as from weapons tests. It is well mixed as $^{14}CO_2$ in the lower atmosphere and oceans, where it is incorporated into all living organisms. Its radioactivity decays with a half-life of 5,730 years. Fossil fuels are formed from organic material, but their ^{14}C component has fully decayed over millions of years. Thus, the CO_2 emitted from the burning of fossil fuels is characterized by its absence of ^{14}C, creating plumes of ^{14}C-depleted air near source regions (Levin et al., 2003; Levin and Rödenbeck, 2008).

The addition of 1 ppm of fossil-fuel CO_2 to a contemporary air mass with a background CO_2 mole fraction of 390 ppm and a $^{14}C/^{12}C$ ratio typical of the free troposphere causes the $^{14}C/^{12}C$ ratio of CO_2 to decrease by 0.27 percent (e.g., Turnbull et al., 2007). Accelerator mass spectrometry can measure ^{14}C in CO_2 in only 2 liters of ambient air to a precision of 0.2 percent. Comparison of the ^{14}C depletion in an air sample to ^{14}C measured in background air determines the CO_2 component from recent fossil-fuel burning to a precision of 0.7 ppm, which is within the range of signals produced by fossil-fuel burning in individual nations (Table B.1, Appendix B).

Global $^{14}CO_2$ observations would greatly reduce transport errors. The fossil-fuel emissions inventories of most Annex I countries are believed to be fairly accurate (Chapter 2). If a baseline of atmospheric $^{14}CO_2$ observations can be established before emission reductions start taking hold, the observations would help constrain atmospheric mixing processes within transport models over the continents.

Box 4.2 presents an analysis of the kind of gains in estimation accuracy that could be realized from systematic ^{14}C measurements if transport errors could be eliminated. The analysis shows that 10,000 $^{14}CO_2$ samples per year could yield fossil fuel CO_2 emission uncertainties of 10-25 percent in the United States. With a better designed sampling grid, only 5,000 samples may be needed to achieve that accuracy. In practice, $^{14}CO_2$ measurements would likely provide

> **BOX 4.2 Gains from ^{14}C Sampling**
>
> A recent Observation System Simulation Experiment (OSSE) tested the capacity of 800 atmospheric $^{14}CO_2$ measurements per month (e.g., Turnbull et al., 2007) to constrain U.S. fossil-fuel CO_2 emissions. The results of the experiment are shown in the figure below. The left panel shows the total flux in 60 regions, as represented by the VULCAN emissions inventory. The right panel shows the uncertainty in the retrieval of the VULCAN emissions, which is treated as truth in this simulation.
>
> Note the change in color scale between the panels. Starting with an assumption of 100 percent uncertainty in the flux from each region, this simulation suggests that $^{14}CO_2$ measurements can potentially determine the fossil-fuel flux in the coterminous United States to better than 10 percent. This OSSE was a test of the sampling density and assumed that the tracer-transport model was perfect. Errors in the tracer-transport model (as discussed here) would increase these uncertainties.
>
>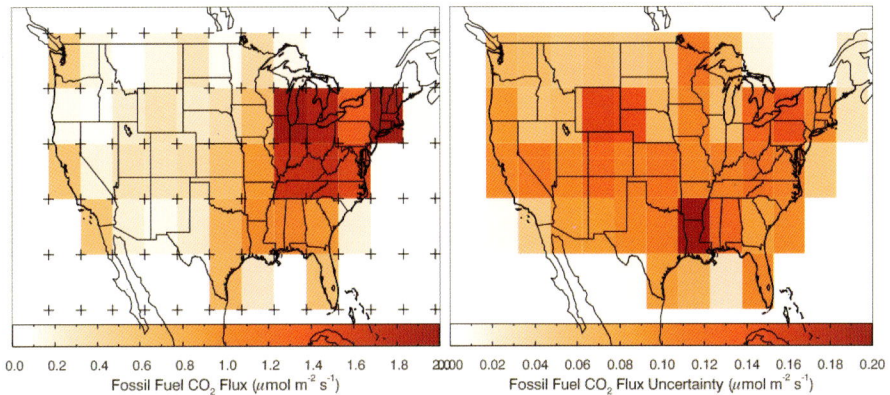
>
> Estimation of CO_2 fluxes from fossil-fuel burning using 10,000 $^{14}CO_2$ measurements per year. *Left panel:* Sources (VULCAN inventory; <www.purdue.edu/eas/carbon/vulcan>) aggregated into 5 × 5 degree areas (colored rectangles). The 84 virtual sampling locations, where virtual $^{14}CO_2$ measurements were made every 3 days at 14:00 local time, are plotted as "pluses." *Right panel:* Uncertainty of the total estimated flux in each 5 × 5 degree area. For both panels, January emissions and uncertainties are shown, but those for other months are not substantially different. SOURCE: Courtesy of John Miller, NOAA/ESRL.

estimates of the global fossil-fuel flux that would be accurate enough to check trends from self-reported inventories collectively (Table 4.1). The accuracy of estimates of fossil-fuel CO_2 emissions by continents would be expected to improve from >100 percent uncertainty to 10-25 percent uncertainty for intensely sampled regions, but national fossil-fuel CO_2 emissions would still be uncertain at the 50-99 percent level if a country is not sampled intensely.

Making $^{14}CO_2$ measurements in the air samples currently being collected in the global cooperative air sampling network (Figure 4.2) would be relatively inexpensive. It would cost only about $5 million to $10 million per year to collect and process the samples and to make the $^{14}CO_2$ measurements on 10,000 samples per year. A dense deployment in the United States (5,000 samples per year) would provide proof of concept and identify its limits.

Measurement Programs to Reduce Model Errors

To improve estimates of national anthropogenic CO_2 emissions from tracer-transport inversion, we will ultimately have to reduce transport errors in atmospheric models (see the plus signs in Table 4.1). Improvement of atmospheric tracer-transport models will require not only improved meteorological observations (especially of vertical mixing) but also new high-resolution measurements of multiple independent tracers that will enable the transport characteristics of the models to be validated. Such data are collected from international networks, supplemented by regional or global campaigns that measure multiple tracers with three-dimensional resolution, including vertical profiles. Previous and ongoing studies of this sort include the TRACE-P (TRAnsport and Chemical Evolution over the Pacific) campaign that examined the emissions of CO from combustion in Asia (Kiley et al., 2003), the INTEX/NA (Intercontinental Chemical Transport Experiment—North America; Choi et al., 2008) that suggested stronger than expected vertical mixing, the regional campaign in the north central United States anchored by the WLEF tall tower in Wisconsin (Denning et al., 2003; Wang et al., 2007), and European efforts (e.g., see Ahmadov et al., 2007; Levin and Karstens, 2007; Vetter et al., 2008). To test and improve tracer-transport models, data on the time evolution of spatially resolved tracer abundances need to be accompanied by high-resolution gridded data on emissions (e.g., Gurney et al., 2009).

Although OCO would be capable of measuring large urban regions or power plants with precisions as great as 1-2 ppm (see Appendix C, Table C.1), much of the atmospheric signal from smaller or dispersed sources is much less than 1 ppm. Such signals can be measured accurately with in situ networks (e.g., surface sites, towers, aircraft profiling), such as the NOAA ESRL network (Figure 4.2) and the World Meteorological Organization (WMO) Global Atmosphere Watch (GAW) stations.[3] However, many areas, including those with high greenhouse gas emissions, are not adequately sampled. For example, CO_2 observations are sparse in the tropical latitudes, Africa, and South America. A number of recent studies have examined how to expand the current CO_2 measurement sites to optimize the ability to retrieve CO_2 emissions from transport modeling (Gloor et al., 2000; Patra et al., 2003; Rayner, 2004; Gurney et al., 2008). Expansion of the GAW network to observe the variations in greenhouse gas abundances in countries with the largest emissions would greatly improve the independent verification of emissions through tracer-transport modeling.

Expansion of the GAW network in the vertical dimension would allow more meaningful comparisons with satellite retrievals of column-averaged CO_2 than ground-based measurements alone and would also provide a routine experimental check on how models transport emissions from the boundary layer to the free troposphere. With the cooperation of commercial airlines, calibrated high-precision measurements can be made routinely from regular flights at low cost. Although the transects are in the upper troposphere and lower stratosphere, a vertical profile is obtained during each take-off and landing. Routine measurements of air pollutants have been made from European commercial airliners as part of the MOZAIC (since 1994) and CARIBIC (since 1997) research programs, and a successor program that will include CO_2 and CH_4 measurements (IAGOS) is being planned.[4] Japanese scientists have obtained flask samples on JAL flights between Tokyo and Melbourne since 1993 and continuous measurements and flask samples onboard five aircraft traveling between Asia, Australia, Europe, and North America since 2005 (Machida et al., 2008).

Improvements to Tracer-Transport Inversion with Carbon Data Assimilation

The accuracy of sources and sinks inferred from variations of the atmospheric mole fraction is intimately tied to the accuracy of winds and mixing characteristics in tracer-transport models. Typically, wind and mixing data are obtained from a meteorological data assimilation system, wherein diverse streams of meteorological observations are merged with a numerical weather prediction (NWP) model to yield, every 3 or 6 hours, the best estimate of the atmospheric state and the uncertainty in the estimate. A reanalysis is undertaken occasionally to assimilate decades of meteorological observations into a single NWP model (e.g., Kalnay et al., 1996; Uppala et al., 2005) and, thus, represents the best source of information about climate variability on diurnal, seasonal, and interannual to decadal time scales. Most tracer-transport models or carbon data assimilation systems employ 3-hourly or 6-hourly atmospheric circulation statistics from a reanalysis.

Carbon data assimilation combines, in principle, emissions inventories and data on land use, greenhouse gas abundances, and meteorology with models of the atmosphere, ecosystems, and oceans into a coherent

---

[3] See <http://www.wmo.int/pages/prog/arep/gaw/gaw_home_en.html>.

[4] See <http://www.iagos.org/> for a description of MOZAIC (Measurement of Ozone and Water Vapour on Airbus in-service Aircraft), CARIBIC (Civil Aircraft for the Regular Investigation of the Atmosphere Based on an Instrument Container), and IAGOS (In-service Aircraft for a Global Observing System).

framework so that emission estimates can be constrained by their consistency with the other observations. Sources and sinks of CO_2 and other greenhouse gases at the surface and in the atmosphere column can be derived, as has been demonstrated by the Observation System Simulation Experiment using CO_2 from OCO (Baker et al., 2008), the two-step assimilation and inversion of AIRS CO_2 (Chevallier et al., 2009; Engelen et al., 2009), and assimilation studies of satellite CO, CH_4, and O_3 using three-dimensional chemical transport models (Ménard et al., 2000; Clerbaux et al., 2001; Stajner et al., 2001; Meirink et al., 2008; Tangborn et al., 2009).

Most existing carbon data assimilation efforts are "univariate" in that they are carried out separately from the meteorological assimilation and neglect the spread in the reanalysis. As a result, consistency between the meteorology and trace gas mole fraction and fluxes is not guaranteed, and the quantifiable uncertainty in the circulation is not propagated into a spread in the trace gas mole fractions and, ultimately, to the sources and sinks. Examples of univariate assimilation projects include efforts to assimilate anticipated CO_2 data from the OCO satellite into atmospheric models (Stajner et al., 2001; Baker et al., 2006b; Miller et al., 2007; Chevallier et al., 2007; Feng et al., 2009), the Oregon and California (ORCA) project, and NOAA's CarbonTracker. The Observation System Simulation Experiments showed that the vast volume of OCO data (8,000,000 observations over 16 days) could yield improvements in CO_2 source-sink estimates over continents with an error reduction of greater than a factor of 2 over land, although the degree of improvement depends on assumptions about errors and biases in the observations and models (Baker et al., 2008).

The ORCA project is a regional example of carbon data assimilation (Law et al., 2004; Quaife et al., 2008; Gökede et al., 2010). A terrestrial carbon flux model uses inputs of Landsat remote sensing data on disturbance history and land cover, and is calibrated with inventory biomass data and seasonal carbon and water fluxes from tower sites (Law et al., 2004). The improved model computes prior estimates of carbon fluxes for the land surface, including fluxes for land use and natural ecosystems. The output of the improved model is then used as input to an atmospheric transport model that links the estimated fluxes to atmospheric CO_2 observations. This modeling component ingests three-dimensional meteorology, various remote sensing products (e.g., fraction of absorbed photosynthetically active radiation), fossil-fuel emission inventories, and time series of atmospheric CO_2 mixing ratio data to produce regional estimates of surface fluxes that are used to evaluate regional representativeness of the prior estimated fluxes. Bayesian inversion is then applied to assign scaling factors that align the surface fluxes with the CO_2 observation time series (Gökede et al., 2010). In this way, the model generates a record of its skill at predicting fluxes from natural ecosystems, agriculture, forestry, and land use, which should improve through time. The intent is to reduce uncertainties in fluxes over space and time, which should improve with increased density of atmospheric CO_2 observations.

A quasi-operational global system is CarbonTracker, which inverts for net CO_2 fluxes for land and ocean regions, with a high-resolution (1 × 1 degree latitude and longitude) zoom region over North America, using the boundary layer CO_2 observing network (Peters et al., 2007). Net fluxes are adjusted each week by linear scaling factors for 209 land regions and 30 ocean regions. Emissions from fossil-fuel burning are held fixed. Initial estimates (before adjustment) are provided by the CASA-GFED2 model for terrestrial fluxes (van der Werf et al., 2006) and by an ocean inverse model for the ocean regions. Vertical profile data of CO_2 from aircraft are used to validate the flux results and the modeled atmospheric transport. Uncertainties in the net CO_2 emissions are estimated by comparing different configurations of the setup. Important limitations are (1) the sparseness of CO_2 observations, which forces the assumption of coherent behavior of ecosystems over large distances; (2) the presumed pattern of fluxes, which dominate the high-resolution results; and (3) the accuracy of atmospheric transport and mixing in the model. In addition, some of the resulting emission patterns reflect primarily the a priori CASA-GFED2 model. However, on the scale of North America, CarbonTracker has consistently found net terrestrial uptake of 0.2 to 0.8 Pg C yr^{-1}, varying year to year, which is not a feature of the CASA-GFED2 model.

The most sophisticated carbon data assimilation

to date is the Global and Regional Earth-System Monitoring Using Satellite and In-situ Data (GEMS) project (Hollingsworth et al., 2008),[5] funded by the European Commission. The central atmospheric model is the European Centre for Medium-Range Weather Forecasts Integrated Forecast System (ECMWF IFS), which includes CO_2 as a model prognostic variable, forced by prescribed surface sources and sinks. Radiance data from AIRS have been assimilated into the ECMWF IFS using the 4DVAR system to constrain the modeled atmospheric CO_2 mole fraction in the upper troposphere. The assimilation consistently improves the meteorology, albeit slightly (McNally et al., 2006), and essentially creates, for the first time, gridded synoptic maps of CO_2 in the midtroposphere (Engelen et al., 2009). The modeled upper-troposphere CO_2 concentrations are then used as "observations" in a standard tracer-inversion step, thus updating the surface sources and sinks (Chevallier et al., 2009). These are promising new techniques, but the AIRS observations are weighted in the midtroposphere, so their ability to differentiate surface emissions from different countries is limited.

A new effort in the United States attempts to carry out a multivariate assimilation of meteorology and CO_2 (Kang, 2009) using a carbon-climate model, so that uncertainties in the air flow and transport estimates are propagated to the trace gas variables, and the trace gas data inform the meteorological state, especially in regions with few wind observations. Early results suggest that the spread in the vertical profile of CO_2 is on the order of 1 ppm due to the uncertainty in meteorological fields alone. With the simultaneous assimilation of AIRS CO_2 observations, the spread in the CO_2 profile grows to 1.5 ppm due to uncertainty in the AIRS observations and the propagation of spread in the meteorology to the CO_2 forecast (Liu et al., 2009). This approach, if combined with accurate CO_2 observations, especially near the surface, could effectively quantify the error due to imperfect knowledge of the meteorology, but leaves the errors due to numerical precision and coding structure of the tracer-transport model.

In addition to developing a fully coupled multivariate atmosphere-land-ocean carbon data assimilation system, it should be feasible to greatly improve several aspects of current data assimilation systems. A high spatial resolution nested subgrid could surround each observation site for a better representation of local meteorology. Multiple transport models and multiple source models of the oceans and terrestrial biosphere could be employed to "internalize" biases between different models as increased (and more realistic) uncertainty estimates. Expanded observation systems would increase the constraints, allowing less dependence on each site. Assumptions of coherent behavior of ecosystems over very large spatial scales could be relaxed because they are being made to compensate for the sparseness of observations.

RECOMMENDATIONS

The remote sensing programs described in Chapter 3 together with initiatives to measure $^{14}CO_2$ and emissions from large local sources would significantly improve our capability to check self-reported emissions of CO_2 (see Table 4.4). In particular, they would allow monitoring of land-use activities responsible for a large share of emissions from agriculture, forestry, and other land use and the significant portion of fossil-fuel emissions from countries that are produced by large local sources. They would also greatly improve our capacity to estimate total fossil-fuel emissions from continents. Implementation of initiatives to improve tracer-transport inversions would improve estimates for total fossil- and non-fossil-fuel emissions at the national level and for time scales ranging from daily to annual, enabling decadal changes to be detected. Specific recommendations include:

• The National Aeronautics and Space Administration (NASA) should build and launch a replacement for the Orbiting Carbon Observatory. A technical advisory group should be convened to assess the optimal orbit and sampling strategies for estimating emissions from large local sources.
• Extend the surface-based atmospheric sampling network to research the atmospheric "domes" of greenhouse gases over a representative sample (e.g., 5-10) of large local emitters, such as cities and power plants. Key goals of the research program would be (1)

---

[5] See also GEMS, <http://gems.ecmwf.int>. GEMS concluded on May 31, 2009, and was replaced by a new project Modelling Atmospheric Composition and Climate (MACC).

TABLE 4.4 Potential Improvements in National Emissions Estimates from Atmospheric and Oceanic Measurements and Models

Gas	Major Sectors or Activities	Current Uncertainty for Annual National Emissions	Possible Improvements in 3-5 Years	Uncertainty of Improved Methods
CO_2	Large local sources (e.g., cities, power plants)	5	CO_2 satellite program, including an OCO replacement, new in situ measurements in cities, and a research program to guide network design and satellite validation	2 (annual) 1 (decadal change)
CO_2	Fossil-fuel combustion	4-5	Improved tracer-transport inversion through new observations ($^{14}CO_2$, additional ground stations, OCO replacement) and data assimilation	1-3 (annual) 1-2 (decadal change)
CO_2	Agriculture, forestry, and other land-use net emissions	5	Improved tracer-transport inversion through new satellite and in situ observations	5
CH_4	Total anthropogenic	3-5	Improved tracer-transport models, new satellite and in situ observations, and improved emission models through research	2-3
N_2O	Total anthropogenic	4-5	Improved tracer-transport and emission models, additional observations	3-5
CFCs, PFCs, HFCs, and SF_6	Industrial processes	4-5	Gridded inventories, improved tracer-transport inversion, and measurement of correlated variations of gases	2-5

NOTES: 1 = <10% uncertainty; 2 = 10-25%; 3 = 25-50%; 4 = 50-100%; 5 = >100% (i.e., cannot be certain if it is a source or sink). Ranges represent emission uncertainties in different countries (e.g., 1-3 means that uncertainties are <10% in some countries, 10-25% in some, and 25-50% in others). Uncertainty levels correspond to 2 standard deviations.

to optimize trend detection using intensive sampling approaches, (2) to develop easily deployable and cost-effective sampling approaches for a globally extensive ground-based network, and (3) to provide a means for validating satellite measurements in these complex and understudied environments.

• Extend the international WMO Global Atmospheric Watch network of in situ sampling stations to fill in underrepresented regions globally, thereby improving national sampling of regional greenhouse gas emissions. Expanding the network to increase collection of vertical profiles of greenhouse gases would constrain atmospheric transport and facilitate interpretation of satellite data. The vertical expansion could be done with the cooperation of commercial aircraft and with balloons flown to higher altitudes. Ideally, all major emitting nations and groups of neighboring smaller nations would participate in the cooperative network. The latter may require financial assistance and capacity building to aid the poorest nations that dominate the most undersampled regions.

• Extend the capability of the existing CO_2 sampling network to measure atmospheric ^{14}C. At least one additional U.S. accelerator mass spectrometry laboratory is needed to handle approximately 10,000 new atmospheric $^{14}CO_2$ samples a year.

• An interagency group, with broad participation from the research community, should design a research program to develop gridded high-resolution data on U.S. fossil-fuel emissions and HFC, CFC, and PFC emissions. An important component of these maps should be uncertainty estimates that can be used directly in data assimilation programs. To support the research, the National Science Foundation (NSF), NOAA, NASA, and the Department of Energy should expand campaigns to sample the time evolution of tracer fields at high resolution as well as studies that use the data to improve transport modeling of tracers. This research would feed into an international initiative to publish gridded estimates for fossil-fuel emissions, as recommended in Chapter 2.

• Develop a state-of-the-art carbon data assimilation system that is coupled and/or synergistic with meteorological, land, and oceanographic data assimilation systems for the United States. This would require new approaches for coupling circulation and biogeochemical models and for deriving biogeochemical properties (and hence surface fluxes) from the observations. It would also require enhanced collaboration among federal agencies with carbon observations, especially between NASA and NOAA, so that the best estimates and the uncertainties in the meteorology become integral components of emission estimation from a replacement OCO.

• Sustain the infrastructure to measure natural sources and sinks on land and in the ocean, which must be separated from the total non-fossil-fuel flux to estimate agriculture, forestry, and other land-use (AFOLU) emissions. This requires sustaining the terrestrial flux network of sites (see Chapter 2)—augmented with high-precision, high-accuracy CO_2 sampling for bottom-up and top-down model calibration—and continued measurements of the oceanic sink (see Appendix C).

References

Achard, F., H.D. Eva, H.-J. Stibig, P. Mayaux, J. Gallego, T. Richards, and J.-P. Malingreau, 2002, Determination of deforestation rates of the world's humid tropical forests, *Science*, **297**, 999-1002.

Achard, F., H.D. Eva, P. Mayaux, H.J. Stibig, and A. Belward, 2004, Improved estimates of net carbon emissions from land cover change in the tropics for the 1990s, *Global Biogeochemical Cycles*, **18**, GB2008, doi:10.1029/2003GB002142.

Achard, F., R. DeFries, H. Eva, M. Hansen, P. Mayaux, and H.-J. Stibig, 2007, Pan-tropical monitoring of deforestation, *Environmental Research Letters*, **2**, 045022, doi: 10.1088/1748-9326/2/4/045022.

Ackerman, K.V., and E.T. Sundquist, 2008, Comparison of two U.S. power-plant carbon dioxide emissions data sets, *Environmental Science and Technology*, **42**, 5688-5693.

Ahmadov, R., C. Gerbig, R. Kretschmer, S. Koerner, B. Neininger, A.J. Dolman, and C. Sarrat, 2007, Mesoscale covariance of transport and CO_2 fluxes: Evidence from observations and simulations using the WRF-VPRM coupled atmosphere-biosphere model, *Journal of Geophysical Research*, **112**, D22107, doi:10.1029/2007JD008552.

Akimoto, H., T. Ohara, J. Kurokawa, and N. Horii, 2006, Verification of energy consumption in China during 1996-2003 by using satellite observational data, *Atmospheric Environment*, **40**, 7663-7667.

Akiyama, H., and H. Tsuruta, 2003, Nitrous oxide, nitric oxide, and nitrogen dioxide fluxes from soils after manure and urea application, *Journal of Environmental Quality*, **32**, 423-431.

Al-Saadi, J., A. Soma, R.B. Pierce, J. Szykman, C. Wiedinmyer, L. Emmonms, S. Kondragunta, X. Zhang, C. Kittaka, T. Schaack, and K. Bowman, 2008, Intercomparison of near-real-time biomass burning emissions estimates constrained by satellite fire data, *Journal of Applied Remote Sensing*, **2**, 021504, doi:10.1117/1.2948785.

Asner, G.P., D.E. Knapp, E.N. Broadbent, P.J.C. Oliveira, M. Keller, and J.N. Silva, 2005, Selective logging in the Brazilian Amazon, *Science*, **310**, 480-482.

Asner, G.P., R.F. Hughs, T.A. Varga, D.E. Knapp, and T. Kennedy-Bowdoin, 2009, Environmental and biotic controls over aboveground biomass throughout a tropical rain forest, *Ecosystems*, **12**, 261-278.

Australian Government Department of Climate Change, 2008, Annex 7: Uncertainty analysis, in *Australia's National Greenhouse Accounts: National Inventory Report 2006—Volume 2, The Australia Government Submission to the UN Framework Convention on Climate Change June*, pp. 209-221, available at <http://unfccc.int/national_reports/annex_i_ghg_inventories/national_inventories_submissions/items/4303.php>.

Bacastow, R.B., 1976, Modulation of atmospheric carbon dioxide by the Southern Oscillation, *Nature*, **261**, 116-118.

Baker, D.F., R.M. Law, K.R. Gurney, P. Rayner, P. Peylin, A.S. Denning, P. Bousquet, L. Bruhwiler, Y.-H. Chen, P. Ciais, I.Y. Fung, M. Heimann, J. John, T. Maki, S. Maksyutov, K. Masarie, M. Prather, B. Pak, S. Taguchi, and Z. Zhu, 2006a, TransCom3 inversion intercomparison: Interannual variability of regional CO_2 sources and sinks, 1988-2003, *Global Biogeochemical Cycles*, **20**, GB1002, doi:10.1029/2004GB002439.

Baker, D.F., S.C. Doney, and D.S. Schimel, 2006b, Variational data assimilation for atmospheric CO_2, *Tellus, Series B*, **58**, 359-365.

Baker, D.F., H. Bösch, S.C. Doney, and D.S. Schimel, 2008, Carbon source/sink information provided by column CO_2 measurements from the Orbiting Carbon Observatory, *Atmospheric Chemistry and Physics Discussions*, **8**, 20,051-20,112.

Banfi, S., M. Filippini, and L.C. Hunt, 2003, Fuel tourism in border regions, CEPE Working Paper 23, Center for Energy Policy and Economics, Swiss Federal Institutes of Technology, Zurich, Switzerland, 13 pp.

Barford, C.C., S.C. Wofsy, M.L. Goulden, J.W. Munger, E. Pyle, S.P. Urbanski, L. Hutyra, S.R. Saleska, D. Fitzjarrald, and K. Moor, 2001, Factors controlling long- and short-term sequestration of atmospheric CO_2 in a mid-latitude forest, *Science*, **294**, 1688-1691.

Battle, M., M.L. Bender, P.P. Tans, J.W.C. White, J.T. Ellis, T. Conway, and R.J. Francey, 2000, Global carbon sinks and their variability inferred from atmospheric O_2 and $\delta^{13}C$, *Science*, **287**, 2467-2470.

Bergamaschi, P., C. Frankenberg, J.F. Meirink, M. Krol, F. Dentener, T. Wagner, U. Platt, J.O. Kaplan, S. Körner, M. Heimann, E.J. Dlugokencky, and A. Goede, 2007, Satellite chartography of atmospheric methane from SCIAMACHY on board ENVISAT:

2. Comparison with inverse model simulations, *Journal of Geophysical Research*, **112**, D02304, doi:10.1029/2006JD007268.

Bertram, T.H., A. Heckel, A. Richter, J.P. Burrows, and R.C. Cohen, 2005, Satellite measurements of daily variations in soil NO_x emissions, *Geophysical Research Letters*, **32**, L24812, doi:10.1029/2005GL024640.

Biradar, C.M., P.S. Thenkabail, A. Platonov, X. Xiao, R. Geerken, P. Noojipady, H. Turral, and J. Vithanage, 2008, Water productivity mapping methods using remote sensing, *Journal of Applied Remote Sensing*, **2**, 023544.

Birdsey, R., N. Bates, M. Behrenfield, K. Davis, S.C. Doney, R. Feely, D. Hansell, L. Heath, E. Kasischke, H. Kheshgi, B.E. Law, C. Lee, A.D. McGuire, P. Raymond, and C.J. Tucker, 2009, Carbon cycle observations: Gaps threaten climate mitigation policies, *EOS, Transactions of the American Geophysical Union*, **90**, 292.

Blackard, J.A., M.V., Finco, E.H. Helmer, G.R. Holden, M.L. Hoppus, D.M. Jacobs, A.J. Lister, G.G. Moisen, M.D. Nelson, R. Riemann, B. Ruefenacht, D. Salajanu, D.L. Weyermann, K.C. Winterberger, T.J. Brandeis, R.L. Czaplewski, R.E. McRoberts, P.L. Patterson, and R.P. Tycio, 2008, Mapping U.S. forest biomass using national forest inventory data and moderate resolution information, *Remote Sensing of Environment*, **112**, 1658-1677.

Böttcher, H., K. Eisbrenner, S. Fritz, G. Kindermann, F. Kraxner, I. McCallum, and M. Obersteiner, 2009, An assessment of monitoring requirements and costs of "Reduced Emissions from Deforestation and Degradation," *Carbon Balance and Management*, **4**, doi:10.1186/1750-0680-4-7.

Boumans, R., R. Costanza, J. Farley, M.A. Wilson, R. Portela, J. Rotmans, F. Villa, and M. Grasso, 2002, Modeling the dynamics of the integrated Earth system and the value of global ecosystem services using the GUMBO model, *Ecological Economics*, **41**, 529-560.

Bouwman, A.F., K.W. Vanderhoek, and J.G.J. Olivier, 1995, Uncertainties in the global source distribution of nitrous oxide, *Journal of Geophysical Research, Atmosphere*, **100**, 2785-2800.

Bouwman, A.F., L.J.M. Boumans, and N.H. Batjes, 2002, Modeling global annual N_2O and NO emissions from fertilized fields, *Global Biogeochemical Cycles*, **16**, 1080, doi:1010.1029/2001gb001812.

Breidenich, C., and D. Bodansky, 2009, *Measurement, Reporting and Verification in a Post-2012 Climate Agreement*, Pew Center on Global Climate Change, Arlington, VA, 32 pp., available at <http://www.pewclimate.org/docUploads/mrv-report.pdf>.

Campbell, J.L., D.C. Donato, D.A. Azuma, and B.E. Law, 2007, Pyrogenic carbon emission from a large wildfire in Oregon, USA, *Journal of Geophysical Research*, **112**, G04014, doi:10.1029/2007JG000451.

Canadell, J.G., C. Le Quéré, M.R. Raupach, C.B. Field, E.T. Buitenhuis, P. Ciais, T.J. Conway, N.P. Gillett, R.A. Houghton, and G. Marland, 2007, Contributions to accelerating atmospheric CO_2 growth from economic activity, carbon intensity, and efficiency of natural sinks, *Proceedings of the National Academy of Sciences*, **47**, 18,866-18,870.

CCSP (Climate Change Science Program), 2007, *The First State of the Carbon Cycle Report (SOCCR): The North American Carbon Budget and Implications for the Global Carbon Cycle*, A.W. King, L. Dilling, G.P. Zimmerman, D.M. Fairman, R.A. Houghton, G. Marland, A.Z. Rose, and T.J. Wilbanks, eds., U.S. Climate Change Science Program and the Subcommittee on Global Change Research, Asheville, N.C., 242 pp.

Chevallier, F., F.-M. Bréon, and P.J. Rayner, 2007, Contribution of the Orbiting Carbon Observatory to the estimation of CO_2 sources and sinks: Theoretical study in a variational data assimilation framework, *Journal of Geophysical Research*, **112**, D09307, doi:10.1029/2006JD007375.

Chevallier, F., R.J. Engelen, C. Carouge, T.J. Conway, P. Peylin, C. Pickett-Heaps, M. Ramonet, P.J. Rayner, and I. Xueref-Remy, 2009, AIRS-based versus flask-based estimation of carbon surface fluxes, *Journal of Geophysical Research*, **114**, D20303, doi:10.1029/2009JD012311.

Choi, Y., S.A. Vay, K.P. Vadrevu, A.J. Soja, J.-H. Woo, S.R. Nolf, G.W. Sachse, G.S. Diskin, D.R. Blake, N.J. Blake, H.B. Singh, M.A. Avery, A. Fried, L. Pfister, and H.E. Fuelberg, 2008, Characteristics of the atmospheric CO_2 signal as observed over the conterminous United States during INTEX-NA, *Journal of Geophysical Research*, **113**, D07301, doi:10.1029/2007JD008899.

Ciais, P., P.P. Tans, J.W.C. White, M. Trolier, R.J. Francey, J.A. Berry, D.R. Randall, P.J. Sellers, J.G. Collatz, and D.S. Schimel, 1995, Partitioning of ocean and land uptake of CO_2 as inferred by $\delta^{13}C$ measurements from the NOAA/CMDL global air sampling network, *Journal of Geophysical Research*, **100**, 5051-5070.

Clerbaux, C., and D.M. Cunnold, 2006, Long-lived compounds, in *Scientific Assessment of Ozone Depletion: 2006*, World Meteorological Organization, Report No. 50, Geneva, Switzerland, 572 pp.

Clerbaux, C., J. Hadji-Lazaro, D. Hauglustaine, G. Mégie, B. Khattatov, and J.-F. Lamarque, 2001, Assimilation of carbon monoxide measured from satellite in a three-dimensional chemistry-transport model, *Journal of Geophysical Research*, **106**, 15,385-15,394.

Conway, T.J., P.P. Tans, L.S. Waterman, K.W. Thoning, D.R. Kitzis, K.A. Masarie, and N. Zhang, 1994, Evidence for interannual variability of the carbon cycle from the NOAA/GMCC global air sampling network, *Journal of Geophysical Research*, **99**, 22,831-22,855.

Crutzen, P.J., A.R. Mosier, K.A. Smith, and W. Winiwarter, 2008, N_2O release from agro-biofuel production negates global warming reduction by replacing fossil fuels, *Atmospheric Chemistry and Physics*, **8**, 389-395.

Curtis, P.S., P.J. Hanson, P. Bolstad, C. Barford, J.C. Randolph, H.P. Schmid, and K.B. Wilson, 2002, Biometric and eddy-covariance based estimates of annual carbon storage in five eastern North American deciduous forests, *Agricultural and Forest Meteorology*, **113**, 3-19.

Danish National Environmental Research Institute, 2008, Annex 7: Table 6.1 and 6.2 of the IPCC good practice guidance, in *Denmark's National Inventory Report 2008: Emission Inventories 1990-2006—Submitted under the United Nations Framework Convention on Climate Change*, pp. 681-685, available at <http://unfccc.int/national_reports/annex_i_ghg_inventories/national_inventories_submissions/items/4303.php>.

Daughtry, C.S.T., P.C. Doraiswamy, E.R. Hunt Jr., A.J. Stern, J.E. McMurtrey III, and J.H. Prueger, 2006, Remote sensing of crop residue cover and soil tillage intensity, *Soil and Tillage Research*, **91**, 101-108.

Davidson, E.A., 2009, The contribution of manure and fertilizer nitrogen to atmospheric nitrous oxide since 1860, *Nature Geoscience*, **2**, 659-662.

Deay, D.S., K.A. Smith, and A.C. Edwards, 2003, Nitrous oxide emission from agricultural drainage waters, *Global Change Biology*, **9**, 195-203.

DeFries, R., F. Achard, F. Brown, M. Herold, D. Murdiyarso, B. Schlamadinger, and C.M. Souza, 2007, Earth observations for estimating greenhouse gas emissions from deforestation in developing countries, *Environmental Science and Policy*, **10**, 385-394.

Denman, K.L., G. Brasseur, A. Chidthaisong, P. Ciais, P.M. Cox, R.E. Dickinson, D. Hauglustaine, C. Heinze, E. Holland, D. Jacob, U. Lohmann, S. Ramachandran, P.L. da Silva Dias, S.C. Wofsy, and X. Zhang, 2007, Couplings between changes in the climate system and biogeochemistry, in *Climate Change 2007: The Physical Science Basis*, Contribution of Working Group I to the Fourth Assessment Report of the Intergovernmental Panel on Climate Change, S. Solomon, D. Qin, M. Manning, Z. Chen, M. Marquis, K.B. Averyt, M. Tignor, and H.L. Miller, eds., Cambridge University Press, New York, pp. 499-587.

Denmead, O.T., 2008, Approaches to measuring fluxes of methane and nitrous oxide between landscapes and the atmosphere, *Plant and Soil*, **309**, 5-24.

Denning, A.S., M. Nicholls, L. Prihodko, I. Baker, P.-L. Vidale, K. Davis, and P. Bakwin, 2003, Simulated and observed variations in atmospheric CO_2 over a Wisconsin forest, *Global Change Biology*, **9**, 1241-1250.

Desjardins, R.L., M.C. McBain, S.K. Kaharabata, E. Pattey, I.J. MacPherson, D.E. Worth, R. Srinivasan, X.P.C. Vergé, and Z. Gao, 2007, Measuring methane emissions from agricultural sources, *International Journal of Applied Environmental Sciences*, **2**, 31-39.

Dhakal, S., S. Kaneko, and H. Imura, 2003, CO_2 emissions from energy use in East-Asian mega-cities, in *Proceedings of the International Workshop on Policy Integration Towards Sustainable Urban Use for Cities in Asia*, February 4-5, East-West Center, Honolulu, Hawaii, available at <http://enviroscope.iges.or.jp/contents/6/index.html>.

Diuk-Wasser, M.A., G. Dolo, M. Bagayokos, N. Sogoba, M.B. Toure, M. Moghaddam, N. Manoukis, S. Rian, S.F. Traore, and C.E. Taylor, 2006, Patterns of irrigated rice growth and malaria vector breeding in Mali using multi-temporal ERS-2 synthetic aperture radar, *International Journal of Remote Sensing*, **27**, 535-548.

Dlugokencky, E.J., B.P. Walter, K.A. Masarie, P.M. Lang, and E.S. Kasischke, 2001, Measurements of an anomalous global methane increase during 1998, *Geophysical Research Letters*, **28**, 499-502.

Drake, J.B., R.O. Dubayaha, D.B. Clark, R.G. Knox, J.B. Blair, M.A. Hofton, R.L. Chazdone, J.F. Weishampel, and S.D. Prince, 2002, Estimation of tropical forest structural characteristics using large-footprint lidar, *Remote Sensing of Environment*, **79**, 305-319.

Engelen, R.J., A.S. Denning, and K.R. Gurney, 2002, On error estimation in atmospheric CO_2 inversions, *Journal of Geophysical Research*, **107**, 4635, doi:10.1029/2002JD002195.

Engelen, R.J., S. Serrar, and F. Chevallier, 2009, Four-dimensional data assimilation of atmospheric CO_2 using AIRS observations, *Journal of Geophysical Research*, **114**, D03303, doi:10.1029/2008JD010739.

Enting, I.G., 2002, *Inverse Problems in Atmospheric Constituent Transport*, Cambridge University Press, Cambridge, U.K., 392 pp.

EPA (Environmental Protection Agency), 2005, *Plain English Guide to the Part 75 Rule*, Environmental Protection Agency, Washington D.C., 129 pp.

EPA, 2008, *Inventory of U.S. Greenhouse Gas Emissions and Sinks: 1990-2006*, EPA 430-R-08-005, Office of Atmospheric Programs, Washington, D.C., available at <http://www.epa.gov/climatechange/emissions/usgginventory.html>.

Feng, L., P.I. Palmer, H. Bösch, and S. Dance, 2009, Estimating surface CO_2 fluxes from space-borne CO_2 dry air mole fraction observations using an ensemble Kalman Filter, *Atmospheric Chemistry and Physics*, **9**, 2619-2633.

Finnigan, J.J., R. Clement, Y. Malhi, R. Leuning, and H.A. Cleugh, 2003, A re-evaluation of long-term flux measurement techniques part I: Averaging and coordinate rotation, *Boundary Layer Meteorology*, **107**, 1-48.

Fletcher, S.E.M., P.P. Tans, L.M. Bruhwiler, J.B. Miller, and M. Heimann, 2004, CH_4 sources estimated from atmospheric observations of CH_4 and its $^{13}C/^{12}C$ isotopic ratios: 2. Inverse modeling of CH_4 fluxes from geographical regions, *Global Biogeochemical Cycles*, **18**, GB4005, doi:10.1029/2004GB002224.

FLIIWG (Future of Land Imaging Interagency Working Group), 2007, *A Plan for a U.S. National Land Imaging Program*, National Science and Technology Council, 110 pp., available at <http://www.landimaging.gov/fli_iwg_report_print_ready_low_res.pdf>.

Forster, P., V. Ramaswamy, P. Artaxo, T. Berntsen, R. Betts, D.W. Fahey, J. Haywood, J. Lean, D.C. Lowe, G. Myhre, J. Nganga, R. Prinn, G. Raga, M. Schulz, and R. Van Dorland, 2007, Changes in atmospheric constituents and in radiative forcing, in *Climate Change 2007: The Physical Science Basis*, Contribution of Working Group I to the Fourth Assessment Report of the Intergovernmental Panel on Climate Change, S. Solomon, D. Qin, M. Manning, Z. Chen, M. Marquis, K.B. Averyt, M. Tignor, and H.L. Miller, eds., Cambridge University Press, New York, pp. 129-234.

Fowler, D., K. Pilegaard, M.A. Sutton, and 57 coauthors, 2009, Atmospheric composition change: Ecosystems-atmosphere interactions, *Atmospheric Environment*, **43**, 5193-5267.

Francey, R.J., P.P. Tans, C.E. Allison, I.G. Enting, J.W.C. White, and M. Trolier, 1995, Changes in oceanic and terrestrial carbon uptake since 1982, *Nature*, **373**, 326-330.

Friedl, M.A., D.K. McIver, J.C.F. Hodges, X.Y. Zhang, D. Muchoney, A.H. Strahler, C.E. Woodcock, S. Gopal, A. Schneider, A. Cooper, A. Baccini, F. Gao, and C. Schaaf, 2002, Global land cover mapping from MODIS: Algorithms and early results, *Remote Sensing of Environment*, **83**, 287-302.

Friedlingstein, P., P. Cox, R. Betts, L. Bopp, W. Von Bloh, V. Brovkin, P. Cadule, S. Doney, M. Eby, I. Fung, G. Bala, J. John, C. Jones, F. Joos, T. Kato, M. Kawamiya, W. Knorr, K. Lindsay, H.D. Matthews, T. Raddatz, P. Rayner, C. Reick, E. Roeckner, K.G. Schnitzler, R. Schnur, K. Strassmann, A.J. Weaver, C. Yoshikawa, and N. Zeng, 2006, Climate-carbon cycle feedback analysis: Results from the (CMIP)-M-4 model intercomparison, *Journal of Climate*, **19**, 3337-3353.

Frolking, S., C. Li, R. Braswell, and J. Fuglestvedt, 2004, Short- and long-term greenhouse gas and radiative forcing impacts of changing water management in Asian rice paddies, *Global Change Biology*, **10**, 1180-1196.

Fung, I., J. Lerner, E. Matthews, M. Prather, L.P. Steele, and P.J. Fraser, 1991, 3-dimensional model synthesis of the global methane cycle, *Journal of Geophysical Research, Atmosphere*, **96**, 13,033-13,065.

Galloway, J.N., A.R. Townsend, J.W. Erisman, M. Bekunda, Z. Cai, J.R. Freney, L.A. Martinelli, S.P. Seitzinger, and M.A. Sutton, 2008, Transformation of the nitrogen cycle: Recent trend, questions, and potential solutions, *Science*, **320**, 889-892.

Geller, L.S., J.W. Elkins, J.M. Lobert, A. Clarke, D.F. Hurst, J.H. Butler, and R.C. Myers, 1997, Tropospheric SF_6: Observed latitudinal distribution and trends, derived emissions and interhemispheric exchange time, *Geophysical Research Letters*, **24**, 675-678.

German Federal Environmental Agency (unweltbundesamt), 2008, Anhang 7: Table 6.1 of the IPCC good practice guidance, in *Submission Under the United Nations Framework Convention on Climate Change 2008: National Inventory Report for the German Greenhouse Gas Inventory 1990-2006*, pp. 519-525, available at <http://unfccc.int/national_reports/annex_i_ghg_inventories/national_inventories_submissions/items/4303.php>.

Giglio, L., G.R. van der Werf, J.T. Randerson, G.J. Collatz, and P. Kasibhatla, 2006, Global estimation of burned area using MODIS active fire observations, *Atmospheric Chemistry and Physics*, **6**, 957-974.

Giglio, L., T. Loboda, D.P. Roy, B. Quayle, and C.O. Justice, 2009, An active-fire based burned area mapping algorithm for the MODIS sensor, *Remote Sensing of Environment*, **113**, 408-420.

Gloor, M., S.M. Fan, S. Pacala, and J. Sarmiento, 2000, Optimal sampling of the atmosphere for purpose of inverse modeling: A model study, *Global Biogeochemical Cycles*, **14**, 407-428.

GOFC-GOLD (Global Observations of Forest and Land Cover Dynamics), 2008, Reducing greenhouse gas emissions from deforestation and degradation in developing countries: A sourcebook of methods and procedures for monitoring, measuring and reporting, GOFC-GOLD Report version COP13-2, Natural Resources Canada, Alberta, available at <http://www.gofc-gold.uni-jena.de/redd/>.

Gökede, M., A.M. Michalak, D. Vickers, D.P. Turner, and B.E. Law, 2010, Atmospheric inverse modeling to constrain regional scale CO_2 budgets at high spatial and temporal resolution, *Geophysical Research Letters*, in press.

Gómez, D.R., J.D. Watterson, B.B. Americano, C. Ha, G. Marland, E. Matsika, L.N. Namayanga, B. Osman-Elasha, J.D. Kalenga Saka, and K. Treanton, 2006, Stationary combustion, in *Volume 2: Energy, 2006 IPCC Guidelines for National Greenhouse Gas Inventories*, H.S. Eggleston, L. Buendia, K. Miwa, T. Ngara, and K. Tanabe, eds., Institute for Global Environmental Strategies, Hayama, Kanagawa, Japan, pp. 2.1-2.47.

Goodale, C.L., M.J. Apps, R.A. Birdsey, C.B. Field, L.S. Heath, R.A. Houghton, J.C. Jenkins, G.H. Kohlmaier, W. Kurz, S. Liu, G.-J. Nabuurs, S. Nilsson, and A.Z. Shvidenko, 2002, Forest carbon sinks in the northern hemisphere, *Ecological Applications*, **12**, 891-899.

Goward, S.N., J. Masek, W. Cohen, G. Moisen, G. Collatz, S. Healey, R. Houghton, C. Huang, R. Kennedy, B. Law, S. Powell, D. Turner, and M.A. Wulder, 2008, Forest disturbance and North American carbon flux, *EOS, Transactions of the American Geophysical Union*, **89**, 105-116.

Grainger, A., 2008, Difficulties in tracking the long-term global trend in tropical forest area, *Proceedings of the National Academy of Sciences*, **105**, 818-823.

Greek Ministry for the Environment, Physical Planning and Public Works, 2008, Annex IV: Uncertainty analysis, in *Climate Change Emissions Inventory: Annual Inventory Submission Under the Convention and the Kyoto Protocol for Greenhouse and Other Gases for the Years 1990-2006*, pp. 255-261, available at <http://unfccc.int/national_reports/annex_i_ghg_inventories/national_inventories_submissions/items/4303.php>.

Gregg, J.S., R.J. Andres, and G. Marland, 2008, China: Emissions pattern of the world leader in CO_2 emissions from fossil fuel consumption and cement production, *Geophysical Research Letters*, **35**, L08806, doi:10.1029/2007GL032887.

Gruber, N., M. Gloor, S.E. Mikaloff Fletcher, S.C. Doney, S. Dutkiewicz, M.J. Follows, M. Gerber, A.R. Jacobson, F. Joos, K. Lindsay, D. Menemenlis, A. Mouchet, S.A. Müller, J.L. Sarmiento, and T. Takahashi, 2009, Oceanic sources, sinks, and transport of atmospheric CO_2, *Global Biogeochemical Cycles*, **23**, GB1005, doi:10.1029/2008GB003349.

Gurney, K.R., R.M. Law, A.S. Denning, P.J. Rayner, D. Baker, P. Bousquet, L. Bruhwiler, Y.H. Chen, P. Ciais, S. Fan, I.Y. Fung, M. Gloor, M. Heimann, K. Higuchi, J. John, T. Maki, S. Maksyutov, K. Masarie, P. Peylin, M. Prather, B.C. Pak, J. Randerson, J. Sarmiento, S. Taguchi, T. Takahashi, and C.W. Yuen, 2002, Towards robust regional estimates of CO_2 sources and sinks using atmospheric transport models, *Nature*, **415**, 626-630.

Gurney, K.R., R.M. Law, A.S. Denning, P.J. Rayner, D. Baker, P. Bousquet, L. Bruhwiler, Y.H. Chen, P. Ciais, S. Fan, I.Y. Fung, M. Gloor, M. Heimann, K. Higuchi, J. John, E. Kowalczyki, T. Maki, S. Maksyutov, P. Peylin, M. Prather, B.C. Pak, J. Sarmiento, S. Taguchi, T. Takahashi, and C.W. Yuen, 2003, Transcom 3 CO_2 inversion intercomparison: 1. Annual mean control results and sensitivity to transport and prior flux information, *Tellus, Series B*, **55**, 555-579.

Gurney, K.R., D. Baker, P. Rayner, and S. Denning, 2008, Interannual variations in continental-scale net carbon exchange and sensitivity to observing networks estimated from atmospheric CO_2 inversions for the period 1980 to 2005, *Global Biogeochemical Cycles*, **22**, GB3025, doi:10.1029/2007GB003082.

Gurney, K.R., D.L. Mendoza, Y. Zhou, M.L. Fischer, C.C. Miller, S. Geethakumar, and S. de la Rue du Can, 2009, High resolution fossil fuel combustion CO_2 emission fluxes for the United States, *Environmental Science and Technology*, **43**, 5535-5541.

Hansen, M.C., S.V. Stehman, P.V. Potapov, T.R. Loveland, J.R.G. Townshend, R.S. DeFries, K.W. Pittman, B. Arunarwati, F. Stolle, M.K. Steininger, M. Carroll, and C. DiMiceli, 2008, Humid tropical forest clearing from 2000 to 2005 quantified by using multitemporal and multiresolution remotely sensed data, *Proceedings of the National Academy of Sciences*, **105**, 9439-9444.

Hawbaker, T.J., V.C. Radeloff, A.D. Syphard, Z.L. Zhu, and S.I. Stewart, 2008, Detection rates of the MODIS active fire product in the United States, *Remote Sensing of Environment*, **112**, 2656-2664.

Hein, R., P.J. Crutzen, and M. Heimann, 1997, An inverse modeling approach to investigate the global atmospheric methane cycle, *Global Biogeochemical Cycles*, **11**, 43-76.

Herold, A., 2007, *Comparison of Verified CO_2 Emissions Under the EU Emission Trading Scheme with National Greenhouse Gas*

Inventories for the Year 2005, European Topic Center on Air and Climate Change, Technical Paper 2007/3, 81 pp., available at <http://air-climate.eionet.europa.eu/reports/ETCACC_TechnPaper_2007_3_CO2_ETS_vs_GHGinv2005>.

Hirsch, A.I., A.M. Michalak, L.M. Bruhwiler, W. Peters, E.J. Dlugokencky, and P.P. Tans, 2006, Inverse modeling estimates of the global nitrous oxide surface flux from 1998-2001, *Global Biogeochemical Cycles*, **20**, GB1008, doi:10.1029/2004GB002443.

Höhne, N., and J. Harnisch, 2002, Comparison of emissions estimates derived from atmospheric measurements with national estimates of HFCs, PFCs and SF_6, in *Non-CO_2 Greenhouse Gases: Scientific Understanding, Control Options and Policy Aspects*, J. Van Ham, A.P.M. Baede, R. Guicherit, and J.G.F.M. Williams-Jacobse, eds., Milpress, Rotterdam, pp. 547-552.

Hollingsworth, A., R.J. Engelen, C. Textor, A. Benedetti, O. Boucher, F. Chevallier, A. Dethof, H. Elbern, H. Eskes, J. Flemming, C. Granier, J.W. Kaiser, J.-J. Morcrette, P. Rayner, V.-H. Peuch, L. Rouil, M.G. Schultz, A.J. Simmons, and the GEMS Consortium, 2008, Toward a monitoring and forecasting system for atmospheric composition: The GEMS Project, *Bulletin of the American Meteorological Society*, **89**, 1147-1164.

Houghton, A., 2003, Revised estimates of the annual net flux of carbon from changes in land use and land management 1850-2000, *Tellus, Series B*, **55**, 378-390.

Houghton, R.A., K.T. Lawrence, J.L. Hackler, and S. Brown, 2001, The spatial distribution of forest biomass in the Brazilian Amazon: A comparison of estimates, *Global Change Biology*, **7**, 731-746.

Houghton, R.A., F. Hall, and S.J. Goetz, 2009, Importance of biomass in the global carbon cycle, *Journal of Geophysical Research*, **114**, G00e03, doi:10.1029/2009jg000935.

Houweling, S., T. Kaminski, F. Dentener, J. Lelieveld, and M. Heimann, 1999, Inverse modeling of methane sources and sinks using the adjoint of a global transport model, *Journal of Geophysical Research, Atmosphere*, **104**, 26,137-26,160.

Hsueh, D.Y., N.Y. Krakauer, J.T. Randerson, X. Xu, S.E. Trumbore, and J.R. Southon, 2007, Regional patterns of radiocarbon and fossil fuel-derived CO_2 in surface air across North America, *Geophysical Research Letters*, **34**, L02816, doi:02810.01029/02006gl027032.

Huang, C., S.N. Goward, J.G. Masek, F. Gao, E.F. Vermote, N. Thomas, K. Schleeweis, R.E. Kennedy, Z. Zhu, J.C. Eidenshink, and J.R.G. Townshend, 2009, Development of time series stacks of Landsat images for reconstructing forest disturbance history, *International Journal of Digital Earth*, **2**, 195-218.

Huang, J., A. Golombek, R. Prinn, R. Weiss, P. Fraser, P. Simmonds, E.J. Dlugokencky, B. Hall, J. Elkins, P. Steele, R. Langenfelds, P. Krummel, G. Dutton, and L. Porter, 2008, Estimation of regional emissions of nitrous oxide from 1997 to 2005 using multinetwork measurements, a chemical transport model, and an inverse method, *Journal of Geophysical Research*, **113**, D17313, doi:10.1029/2007JD009381.

Hudiburg, T., B.E. Law, D.P. Turner, J. Campbell, D. Donato, and M. Duane, 2009, Carbon dynamics of Oregon and Northern California forests and potential land-based carbon storage, *Ecological Applications*, **19**, 163-180.

Idso, C.D., S.B. Idso, and R.C. Balling Jr., 2001, An intensive two-week study of an urban CO_2 dome in Phoenix, Arizona, USA, *Atmospheric Environment*, **35**, 995-1000.

IEA, 2009, *2006 IPCC Guidelines Versus the Revised 1996 IPCC Guidelines: Implications for Estimates of CO_2 Emission from Fuel Combustion*, March 25 draft, Paris, France, 22 pp.

IPCC (Intergovernmental Panel on Climate Change), 2000, *IPCC Good Practice Guidance and Uncertainty Management in National Greenhouse Gas Inventories*, available at <http://www.ipcc-nggip.iges.or.jp/public/gp/english/>.

IPCC, 2006, *2006 IPCC Guidelines for National Greenhouse Gas Inventories*, H.S. Eggleston, L. Buendia, K. Miwa, T. Ngara, and K. Tanabe, eds., Institute for Global Environmental Strategies, Hayama, Kanagawa, Japan, 5 volumes.

IPCC, 2007a, *Climate Change 2007: The Physical Science Basis, Contribution of Working Group I to the Fourth Assessment Report of the Intergovernmental Panel on Climate Change*, S. Solomon, D. Qin, M. Manning, Z. Chen, M. Marquis, K.B. Averyt, M. Tignor, and H.L. Miller, eds., Cambridge University Press, New York, 996 pp.

IPCC, 2007b, *Climate Change 2007: Mitigation, Contribution of Working Group III to the Fourth Assessment Report of the Intergovernmental Panel on Climate Change*, B. Metz, O.R. Davidson, P.R. Bosch, R. Dave, and L.A. Meyer, eds., Cambridge University Press, New York, 851 pp.

Ito, A., J.E. Penner, M.J. Prather, C.P. de Campos, R.A. Houghton, T. Kato, A.K. Jain, X. Yang, G.C. Hurtt, S. Frolking, M.G. Fearon, L.P. Chini, A. Wang, and D.T. Price, 2008, Can we reconcile differences in estimates of carbon fluxes from land-use change and forestry for the 1990s? *Atmospheric Chemistry and Physics*, **8**, 3291-3310.

Jacobson, A.R., S.E. Mikaloff Fletcher, N. Gruber, J.L. Sarmiento, and M. Gloor, 2007a, A joint atmosphere-ocean inversion for surface fluxes of carbon dioxide: 1. Methods and global-scale fluxes, *Global Biogeochemical Cycles*, **21**, doi:10.1029/2005GB002556.

Jacobson, A.R., S.E. Mikaloff Fletcher, N. Gruber, J.L. Sarmiento, and M. Gloor, 2007b, A joint atmosphere-ocean inversion for surface fluxes of carbon dioxide: 2. Regional results, *Global Biogeochemical Cycles*, **21**, doi:10.1029/2006GB002703.

Jain, M.C., S. Kumar, R. Wassmann, S. Mitra, S.D. Singh, J.P. Singh, R. Singh, A.K. Yadav, and S. Gupta, 2000, Methane emissions from irrigated rice fields in northern India (New Delhi), *Nutrient Cycling in Agroecosystems*, **58**, 75-83.

Jonas, M., and S. Nilsson, 2007, Prior to economic treatment of emissions and their uncertainties under the Kyoto Protocol: Scientific uncertainties that must be kept in mind, *Water, Air and Soil Pollution: Focus*, **7**, 475-482.

Kaheil, Y.H., and I.F. Creed, 2009, Detecting and downscaling wet areas on boreal landscapes, *Geoscience and Remote Sensing Letters*, **6**, 179-183.

Kalnay, E., M. Kanamitsu, R. Kistler, W. Collins, D. Deaven, L. Gandin, M. Iredell, S. Saha, G. White, J. Woollen, Y. Zhu, A. Leetmaa, R. Reynolds, M. Chelliah, W. Ebisuzaki, W. Higgins, J. Janowiak, K.C. Mo, C. Ropelewski, J. Wang, R. Jenne, and D. Joseph, 1996, The NCEP/NCAR 40-year reanalysis project, *Bulletin of the American Meteorological Society*, **77**, 437-471.

Kang, J.-S., 2009, *Carbon Cycle Data Assimilation Using a Coupled Atmosphere-Vegetation Model and the Local Ensemble Transform Kalman Filter*, Ph.D. thesis, University of Maryland, College Park, 142 pp.

Kaynak, B., Y. Hu, R.V. Martin, C.E. Sioris, and A.G. Russell, 2009, Comparison of weekly cycle of NO_2 satellite retrievals and NO_x emission inventories for the continental United States, *Journal of Geophysical Research*, **114**, D05302, doi:10.1029/2008JD010714.

Keeling, C.D., 1961, The concentration and isotopic abundances of carbon dioxide in rural and marine air, *Geochimica et Cosmochimica Acta*, **24**, 277-298.

Keeling, C.D., T.P. Whorf, M. Wahlen, and J. van der Plichtt, 1995, Interannual extremes in the rate of rise of atmospheric carbon dioxide since 1980, *Nature*, **375**, 666-670.

Keeling, R.F., R.P. Najjar, M.L. Bender, and P.P. Tans, 1993, What atmospheric oxygen measurements can tell us about the global carbon cycle, *Global Biogeochemical Cycles*, **7**, 37-67.

Kennedy, R.E., W.B. Cohen, and T.A. Schroeder, 2007, Trajectory-based change detection for automated characterization of forest disturbance dynamics, *Remote Sensing of Environment*, **110**, 370-386.

Kiley, C.M., H.E. Fuelberg, P.I. Palmer, D.J. Allen, G.R. Carmichael, D.J. Jacob, C. Mari, R.B. Pierce, K.E. Pickering, Y.H. Tang, O. Wild, T.D. Fairlie, J.A. Logan, G.W. Sachse, T.K. Shaack, and D.G. Streets, 2003, An intercomparison and evaluation of aircraft-derived and simulated CO from seven chemical transport models during the TRACE-P experiment, *Journal of Geophysical Research*, **108**, 8819, doi:10.1029/2002JD003089.

Kort, E.A., J. Eluszkiewicz, B.B. Stephens, J.B. Miller, C. Gerbig, T. Nehrkorn, B.C. Daube, J.O. Kaplan, S. Houweling, and S.C. Wofsy, 2008, Emissions of CH_4 and N_2O over the United States and Canada based on a receptor oriented modeling framework and COBRA-NA atmospheric observations, *Geophysical Research Letters*, **35**, L18808, doi:10.1029/2008GL034031.

Kurz, W.A., C.C. Dymond, T.M. White, G. Stinson, C.H. Shaw, G.J. Rampley, C. Smyth, B.N. Simpson, E.T. Neilson, J.A. Trofymow, J. Metsaranta, and M.J. Apps, 2009, CBM-CFS3: A model of carbon-dynamics in forestry and land-use change implementing IPCC standards, *Ecological Modeling*, **220**, 480-504.

Lauvaux, T., M. Uliasz, C. Sarrat, F. Chevallier, P. Bousquet, C. Lac, K.J. Davis, P. Ciais, A.S. Denning, and P.J. Rayner, 2008, Mesoscale inversion: First results from the CERES campaign with synthetic data, *Atmospheric Chemistry and Physics*, **8**, 3459-3471.

Law, B.E., D. Turner, J. Campbell, O.J. Sun, S. Van Tuyl, W.D. Ritts, and W.B. Cohen, 2004, Disturbance and climate effects on carbon stocks and fluxes across western Oregon, USA, *Global Change Biology*, **10**, 1429-1444.

Law, B.E., D. Turner, M. Lefsky, J. Campbell, M. Guzy, O. Sun, S. Van Tuyl, and W. Cohen, 2006, Carbon fluxes across regions: Observational constraints at multiple scales, in *Scaling and Uncertainty Analysis in Ecology: Methods and Applications*, J. Wu, B. Jones, H. Li, and O. Loucks, eds., Springer, Dordrecht, The Netherlands, pp. 167-190.

Law, B.E., T. Arkebauer, J.L. Campbell, J. Chen, O. Sun, M. Schwartz, C. van Ingen, and S. Verma, 2008, *Terrestrial Carbon Observations: Protocols for Vegetation Sampling and Data Submission*, Global Terrestrial Observing System Report 55, Food and Agriculture Organization, Rome, 87 pp.

Law, R.M., P.J. Rayner, L.P. Steele, and I.G. Enting, 2003, Data and modelling requirements for CO_2 inversions using high-frequency data, *Tellus, Series B*, **55**, 512-521.

Law, R.M., R.J. Matear, and R.J. Francey, 2008, Comment on "Saturation of the Southern Ocean CO_2 sink due to recent climate change," *Science*, **319**, 570.

Lefsky, M.A., W.B. Cohen, D.J. Harding, G.G. Parker, S.A. Acker, and S.T. Gower, 2002, Lidar remote sensing of above-ground biomass in three biomes, *Global Ecology and Biogeography*, **11**, 393-399.

Le Quéré, C., 2009, Closing the global budget for CO_2, *Global Change*, **74**, 28-31.

Le Quéré, C., C. Rödenbeck, E.T. Buitenhuis, T.J. Conway, R. Langenfelds, A. Gomez, C. Labuschagne, M. Ramonet, T. Nakazawa, N. Metzl, N. Gillett, and M. Heimann, 2007, Saturation of the Southern Ocean CO_2 sink due to recent climate change, *Science*, **316**, 1735-1738.

Le Quéré, C., M.R. Raupach, J.G. Canadell, G. Marland, L. Bopp, P. Ciais, T.J. Conway, S.C. Doney, R. Feely, P. Foster, P. Friedlingstein, K. Gurney, R.A. Houghton, J.I. House, C. Huntingford, P. Levy, M.R. Lomas, J. Majkut, N. Metzl, J.P. Ometto, G.P. Peters, I.C. Prentice, J.T. Randerson, S.W. Running, J.L. Sarmiento, U. Schuster, S. Sitch, T. Takahashi, N. Viovy, G.R. van der Werf, and F.I. Woodward, 2009, Trends in the sources and sinks of carbon dioxide, *Nature Geoscience*, **2**, 831-836.

Levin, I., and U. Karstens, 2007, Inferring high-resolution fossil fuel CO_2 records at continental sites from combined $^{14}CO_2$ and CO observations, *Tellus, Series B*, **59**, 245-250.

Levin, I., and C. Rödenbeck, 2008, Can the envisaged reductions of fossil fuel CO_2 emissions be detected by atmospheric observations? *Naturwissenschaften*, **95**, 203-208.

Levin, I., B. Kromer, M. Schmidt, and H. Sartorius, 2003, A novel approach for independent budgeting of fossil fuel CO_2 over Europe by $^{14}CO_2$ observations, *Geophysical Research Letters*, **30**, 2194, doi:2110.1029/2003gl018477.

Li, C., J. Qiu, S. Frolking, X. Xiao, W. Salas, B. Moore, S. Boles, Y. Huang, and R. Sass, 2002, Reduced methane emissions from large-scale changes in water management of China's rice paddies during 1980-2000, *Geophysical Research Letters*, **29**, 1972, doi:10.1029/2002GL015370.

Li, C., S. Frolking, X. Xiao, B. Moore III, S. Boles, J. Qiu, Y. Huang, W. Salas, and R. Sass, 2005, Modeling impacts of farming management alternatives on CO_2, CH_4, and N_2O emissions: A case study for water management of rice agriculture of China, *Global Biogeochemical Cycles*, **19**, GB3010.

Liu, J., M. Chahine, I. Fung, E. Kalnay, and E. Olsen, 2009, AIRS CO_2 assimilation with an EnKF, *Eos, Transactions of the American Geophysical Union*, **90**, Fall Meeting Supplement, Abstract A51A-0086.

Lu, D., M. Batistella, P. Mausel, and E. Moran, 2007, Mapping and monitoring land degradation risks in the western Brazilian Amazon using multitemporal Landsat TM/ETM+ images, *Land Degradation and Development*, **18**, 41-54.

Luyssaert, S., E.-D. Schulze, A. Börner, A. Knohl, D. Hessenmöller, B.E. Law, P. Ciais, and J. Grace, 2008, Old-growth forests as global carbon sinks, *Nature*, **455**, 213-215.

Luyssaert, S., P. Ciais, S.L. Piao, E.-D. Schulze, M. Jung, S. Zaehle, M.J. Schelhaas, M. Reichstein, G. Churkina, D. Papale, G. Abril, C. Beer, J. Grace, D. Loustau, G. Matteucci, F. Magnani, G.J. Nabuurs, H. Verbeeck, M. Sulkava, G.R. van der Werf, I.A. Janssens, and members of the CarboEurope-IP synthesis team, 2009, The European carbon balance: Part 3: Forests, *Global Change Biology*, **16**, 1429-1450.

Machida, T., H. Matsueda, Y. Sawa, Y. Nakagawa, K. Hirotani, N. Kondo, K. Goto, T. Nakazawa, K. Ishikawa, and T. Ogawa, 2008, Worldwide measurements of atmospheric CO_2 and other trace gas species using commercial airlines, *Journal of Atmospheric and Oceanic Technology*, **25**, 1744-1754.

Manjunath, K.R., S. Panigrahy, K. Kumari, T.K. Adhya, and J.S. Parihar, 2006, Spatiotemporal modeling of methane flux from the rice fields of India using remote sensing and GIS, *International Journal of Remote Sensing*, **27**, 4701-4707.

Manning, A.C., and R.F. Keeling, 2006, Global oceanic and land biotic carbon sinks from the Scripps atmospheric oxygen flask sampling network, *Tellus, Series B*, **58**, 95-116.

Manning, A.J., D.B. Ryall, R.G. Derwent, P.G. Simmonds, and S. O'Doherty, 2003, Estimating European emissions of ozone depleting and greenhouse gases using observations and a modeling back-attribution technique, *Journal of Geophysical Research*, **108**, 4405, doi:10.1029/2002JD002312.

Marland, G., and B. Schlamadinger, 1997, Forests for carbon sequestration or fossil-fuel substitution? A sensitivity analysis, *Biomass and Bioenergy*, **13**, 389-397.

Marland, G., A. Brenkert, and J. Olivier, 1999, CO_2 from fossil fuel burning: A comparison of ORNL and EDGAR estimates of national emissions, *Environmental Science and Policy*, **2**, 265-273.

Marland, G., R.J. Andres, T.J. Blasing, T.A. Boden, C.T. Broniak, J.S. Gregg, L.M. Losey, and K. Treanton, 2007, Energy, industry, and waste management activities: An introduction to CO_2 emissions from fossil fuels, in *The First State of the Carbon Cycle Report (SOCCR): The North American Carbon Budget and Implications for the Global Carbon Cycle*, A.W. King, L. Dilling, G.P. Zimmerman, D.M. Fairman, R.A. Houghton, G. Marland, A.Z. Rose, and T.J. Wilbanks, eds., U.S. Climate Change Science Program and the Subcommittee on Global Change Research, Asheville, N.C., pp. 57-64.

Marland, G., K. Hamal, and M. Jonas, 2009, How uncertain are estimates of CO_2 emissions? *Journal of Industrial Ecology*, **13**, 4-7.

Martin, R.E., and G.P. Asner, 2005, Regional estimate of nitric oxide emissions following woody encroachment: Linking imaging spectroscopy and field studies, *Ecosystems*, **8**, 33-47.

Martin, R.V., C.E. Sioris, K. Chance, T.B. Ryerson, T.H. Bertram, P.J. Wooldridge, R.C. Cohen, J.A. Neuman, A. Swanson, and F.M. Flocke, 2006, Evaluation of space-based constraints on global nitrogen oxide emissions with regional aircraft measurements over and downwind of eastern North America, *Journal of Geophysical Research*, **111**, D15308, doi:10.1029/2005JD006680.

Masek, J.G., C. Huang, R. Wolfe, K. Cohen, F. Hall, J. Kutler, and P. Nelson, 2008, North American forest disturbance mapped from a decadal Landsat record, *Remote Sensing of Environment*, **112**, 2914-2926.

Mays, K.L., P.B. Shepson, B.H. Stirm, A. Karion, C. Sweeney, and K.R. Gurney, 2009, Aircraft-based measurements of the carbon footprint of Indianapolis, *Environmental Science and Technology*, **43**, 7816-7823.

McMillan, A.M.S., G.C. Winston, and M.L. Goulden, 2008, Age-dependent response of boreal forest to temperature and rainfall variability, *Global Change Biology*, **14**, 1904-1916.

McNally, A.P., P.D. Watts, J.A. Smith, R. Engelen, G.A. Kelly, J.N. Thépaut, and M. Matricardi, 2006, The assimilation of AIRS radiance data at ECMWF, *Quarterly Journal of the Royal Meteorological Society*, **132**, 935-957.

Medvigy, D., S.C. Wofsy, J.W. Munger, D.Y. Hollinger, and P.R. Moorcroft, 2009, Mechanistic scaling of ecosystem function and dynamics in space and time: Ecosystem Demography Model Version 2, *Journal of Geophysical Research*, **114**, G01002, doi:10.1029/2008JG000812.

Meigs, G.W., D.C. Donato, J.L. Campbell, J.G. Martin, and B.E. Law, 2009, Forest fire impacts on carbon uptake, storage, and emission: The role of burn severity in the eastern Cascades, Oregon, *Ecosystems*, **12**, 1246-1267.

Meirink, J.F., P. Bergamaschi, C. Frankenberg, M.T.S. D'Amelio, E.J. Dlugokencky, L.V. Gatti, S. Houweling, J.B. Miller, T. Röckmann, M. Gabriella Villani, and M.C. Krol, 2008, Four-dimensional variational data assimilation for inverse modeling of atmospheric methane emissions: Analysis of SCIAMACHY observations, *Journal of Geophysical Research*, **113**, D17301, doi:10.1029/2007JD009740.

Ménard, R., S.E. Cohn, L.-P. Chang, and P.M. Lyster, 2000, Assimilation of stratospheric chemical tracer observations using a Kalman filter. Part I: Formulation, *Monthly Weather Review*, **128**, 2654-2671.

Mikaloff Fletcher, S.E., P.P. Tans, L.M. Bruhwiler, J.B. Miller, and M. Heimann, 2004, CH_4 sources estimated from atmospheric observations of CH_4 and its $^{13}C/^{12}C$ isotopic ratios: 1. Inverse modeling of source processes, *Global Biogeochemical Cycles*, **18**, GB4004, doi:10.1029/2004GB002223.

Miller, C.E., D. Crisp, P.L. DeCola, S.C. Olsen, J.T. Randerson, A.M. Michalak, A. Alkhaled, P. Rayner, D.J. Jacob, P. Suntharalingam, D.B.A. Jones, A.S. Denning, M.E. Nicholls, S.C. Doney, S. Pawson, H. Bösch, B.J. Connor, I.Y. Fung, D. O'Brien, R.J. Salawitch, S.P. Sander, B. Sen, P. Tans, G.C. Toon, P.O. Wennberg, S.C. Wofsy, Y.L. Yung, and R.M. Law, 2007, Precision requirements for space-based XCO_2 data, *Journal of Geophysical Research*, **112**, D10314, doi:10.1029/2006JD007659.

Miller, J.D., E.E. Knapp, C.H. Key, C.N. Skinner, C.J. Isbell, R.M. Creasy, and J.W. Sherlock, 2009, Calibration and validation of the relative differenced normalized burn ratio (RdNBR) to three measures of fire severity in the Sierra Nevada and Klamath Mountains, California, USA, *Remote Sensing of Environment*, **113**, 645-656.

Morton, D.C., R.S. DeFries, J.T. Randerson, L. Giglio, W. Schroeder, and G.R. van der Werf, 2008, Agricultural intensification increases deforestation fire activity in Amazonia, *Global Change Biology*, **14**, 2262-2275.

Nevison, C.D., T. Lueker, and R.F. Weiss, 2004, Quantifying the nitrous oxide source from coastal upwelling, *Global Biogeochemical Cycles*, **18**, GB1018, doi:10.1029/2003GB002110.

NRC (National Research Council), 2007, *Earth Science and Applications from Space: National Imperatives for the Next Decade and Beyond*, The National Academies Press, Washington, D.C., 456 pp.

Pacala, S.W., G.C. Hurtt, D. Baker, P. Peylin, R.A. Houghton, R.A. Birdsey, L. Heath, E.T. Sundquist, R.F. Stallard, P. Ciais, P. Moorcroft, J.P. Caspersen, E. Shevliakova, B. Moore, G. Kohlmaier, E. Holland, M. Gloor, M.E. Harmon, S.-M. Fan, J.L. Sarmiento, C.L. Goodale, D. Schimel, and C.B. Field, 2001, Consistent land- and atmosphere-based U.S. carbon sink estimates, *Science*, **292**, 2316-2320.

Pacala, S., R.A. Birdsey, S.D. Bridgham, R.T. Conant, K. Davis, B. Hales, R.A. Houghton, J.C. Jenkins, M. Johnston, G. Marland, and K. Paustian, 2007, The North American carbon budget past and present, in *The First State of the Carbon Cycle Report (SOCCR): The North American Carbon Budget and Implications for the Global Carbon Cycle*, A.W. King, L. Dilling, G.P. Zimmerman, D.M. Fairman, R.A. Houghton, G. Marland, A.Z. Rose, and T.J. Wilbanks, eds., Climate Change Science Program and the Subcommittee on Global Change Research, Asheville, N.C., pp. 29-36.

Page, S.E., R.A.J. Wst, D. Weiss, J.O. Rieley, W. Shotyk, and S.H. Limin, 2004, A record of Late Pleistocene and Holocene carbon accumulation and climate change from an equatorial peat bog (Kalimantan, Indonesia): Implications for past, present and future carbon dynamics, *Journal of Quaternary Science*, **19**, 625-635.

Pales, J.C., and C.D. Keeling, 1965, The concentration of atmospheric carbon dioxide in Hawaii, *Journal of Geophysical Research*, **70**, 6053-6077.

Pataki, D.E., D.R. Bowling, and J.R. Ehleringer, 2003, Seasonal cycle of carbon dioxide and its isotopic composition in an urban atmosphere: Anthropogenic and biogenic effects, *Journal of Geophysical Research*, **108**, 4735, doi:10.1029/2003JD003865.

Pataki, D.E., J.T. Randerson, W. Wang, M. Herzenach, and N.E. Grulke, 2010, The isotopic composition of plants and soils as biomarkers of pollution, in *Isoscapes: Understanding Movement, Pattern, and Process on Earth through Isotope Mapping*, J.B. West, G.J. Bowen, T.E. Dawson, and K.P. Tu, eds., Springer, Dordrecht, Germany, pp. 407-423.

Patra, P.K., S. Maksyutov, and Transcom 3 Modelers, 2003, Sensitivity of optimal extension of CO_2 observation networks to model transport, *Tellus, Series B*, **55**, 498-511.

Pattey, E., I.B. Strachan, R.L. Desjardins, G.C. Edwards, D. Dow, and J.I. MacPherson, 2006, Application of a tunable diode laser to the measurement of CH_4 and N_2O fluxes from field to landscape scale using several micrometeorological techniques, *Agricultural and Forest Meteorology*, **136**, 222-236.

Peters, W., A.R. Jacobson, C. Sweeney, A.E. Andrews, T.J. Conway, K. Masarie, J.B. Miller, L.M.P. Bruhwiler, G. Pétron, A.I. Hirsch, D.E.J. Worthy, G.R. van der Werf, J.T. Randerson, P.O. Wennberg, M.C. Krol, and P.P. Tans, 2007, An atmospheric perspective on North American carbon dioxide exchange: CarbonTracker, *Proceedings of the National Academy of Sciences*, **104**, 18,925-18,930.

Peylin, P., D. Baker, J. Sarmiento, P. Ciais, and P. Bousquet, 2002, Influence of transport uncertainty on annual mean and seasonal inversions of atmospheric CO_2 data, *Journal of Geophysical Research*, **107**, 4385, doi:10.1029/2001JD000857.

Phillips, F.A., R. Leuning, R. Baigent, K.B. Kelly, and O.T. Denmead, 2007, Nitrous oxide flux measurements from an intensively managed irrigated pasture using micrometeorological techniques, *Agricultural and Forest Meteorology*, **143**, 92-105.

Poland National Administration of the Emissions Trading Scheme, 2008, Annex 5: Uncertainty assessment of the 2006 inventory, in *Poland's National Inventory Report 2006*, National Emission Centre, available at <http://unfccc.int/national_reports/annex_i_ghg_inventories/national_inventories_submissions/items/4303.php>.

Portuguese Environmental Agency, 2008, Table 2.2—Tier 2 Level Assessment with LULUCF: 1990 and 2006, in *Portuguese National Inventory Report on Greenhouse Gases, 1990-2006—Submitted Under the United Nations Framework Convention on Climate Change and the Kyoto Protocol*, pp. A-568, available at <http://unfccc.int/national_reports/annex_i_ghg_inventories/national_inventories_submissions/items/4303.php>.

Powell, S.L., W.B Cohen, S.P. Healey, R.E. Kennedy, G.G. Moisen, K.B. Pierce, and J.L. Ohmann, 2010, Quantification of live aboveground forest biomass dynamics with Landsat time-series and field inventory data: A comparison of empirical modeling approaches, *Remote Sensing of Environment*, **114**, 1053-1068.

Prather, M., D. Ehhalt, F. Dentener, R.G. Derwent, E. Dlugokencky, E. Holland, I.S.A. Isaksen, J. Katima, V. Kirchhoff, P. Matson, P.M. Midgley, and M. Wang, 2001, Atmospheric chemistry and greenhouse gases, in *Climate Change 2001: The Scientific Basis*, J.T. Houghton, Y. Ding, D.J. Griggs, M. Noguer, P.J. van der Linden, X. Dai, K. Maskell, and C.A. Johnson, eds., Cambridge University Press, New York, pp. 239-287.

Prather, M.J., 1985, Continental sources of halocarbons and nitrous oxide, *Nature*, **317**, 221-225.

Prather, M.J., 1994, Lifetimes and eigenstates in atmospheric chemistry, *Geophysical Research Letters*, **21**, 801-804.

Prather, M.J., and J. Hsu, 2008, NF_3, the greenhouse gas missing from Kyoto, *Geophysical Research Letters*, **35**, L12810, doi:10.1029/2008GL034542.

Prather, M.J., X. Zhu, S.E. Strahan, S.D. Steenrod, and J.M. Rodriguez, 2008, Quantifying errors in trace species transport modeling, *Proceedings of the National Academy of Sciences*, **105**, 19,617-19,621.

Prather, M.J., J.E. Penner, J.S. Fuglestvedt, A. Kurosawa, J.A. Lowe, N. Höhne, A.K. Jain, N. Andronova, L. Pinguelli, C. Pires de Campos, S.C.B. Raper, R.B. Skeie, P.A. Stott, J. van Aardenne, and F. Wagner, 2009, Tracking uncertainties in the causal chain from human activities to climate change, *Geophysical Research Letters*, **36**, L05707, doi:10.1029/2008GL036474.

Prigent, C., F. Papa, F. Aires, W.B. Rossow, and E. Matthews, 2007, Global inundation dynamics inferred from multiple satellite observations, 1993-2000, *Journal of Geophysical Research*, **112**, D12107, doi:10.1029/2006JD007847.

Prinn, R.G., R.F. Weiss, P.J. Fraser, P.G. Simmonds, D.M. Cunnold, S. O'Doherty, P. Salameh, L. Porter, P. Krummel, R.H.J. Wang, B. Miller, C. Harth, B. Greally, F.A. Van Woy, L.P. Steele, J. Muehle, G. Sturrock, F.N. Alyea, J. Huang, and D.E. Hartley, 2005, The ALE/GAGE/AGAGE Network, Carbon Dioxide Information and Analysis Center, doi:10.3334/CDIAC/atg.db1001, available at <http://cdiac.esd.ornl.gov/ndps/alegage.html>.

Quaife, T., P. Lewis, M. De Kauwe, M. Williams, B.E. Law, M. Disney, and P. Bowyer, 2008, Assimilating canopy reflectance data into an ecosystem model with an ensemble Kalman filter, *Remote Sensing of Environment*, **112**, 1347-1364.

Randerson, J.T., G.R. van der Werf, G.J. Collatz, L. Giglio, C.J. Still, P. Kasibhatla, J.B. Miller, J.W.C. White, R.S. DeFries, and E.S. Kasischke, 2005, Fire emissions from C3 and C4 vegetation and their influence on interannual variability of atmospheric CO_2 and $d^{13}CO_2$, *Global Biogeochemical Cycles*, **19**, GB2019, doi:10.1029/2004GB002366.

Raupach, M.R., G. Marland, P. Ciais, C. Le Quéré, J.G. Canadell, G. Klepper, and C.B. Field, 2007, Global and regional drivers of accelerating CO_2 emissions, *Proceedings of the National Academy of Sciences*, **104**, 10,288-10,293.

REFERENCES

Rayner, P.J., 2004, Optimizing CO_2 observing networks in the presence of model error: Results from TransCom 3, *Atmospheric Chemistry and Physics*, **4**, 413-421.

Rayner, P.J., and D.M. O'Brien, 2001, The utility of remotely sensed CO_2 concentration data in surface source inversions, *Geophysical Research Letters*, **28**, 175-178.

Revelle, R., and H.E. Suess, 1957, Carbon dioxide exchange between atmosphere and ocean and the question of an increase of atmospheric CO_2 during past decades, *Tellus*, **9**, 18-27.

Richards, G.P., 2001, The FullCAM Carbon Accounting Model: Development, calibration and implementation for the National Carbon Accounting System, *National Carbon Accounting System Technical Report 28*, Australian Greenhouse Office, Canberra, Australia, 50 pp.

Rigby, M., R. Toumi, R. Fisher, D. Lowry, and E.G. Nisbet, 2008a, First continuous measurements of CO_2 mixing ratio in central London using a compact diffusion probe, *Atmospheric Environment*, **42**, 8943-8953.

Rigby, M., R.G. Prinn, P.J. Fraser, P.G. Simmonds, R.L. Langenfelds, J. Huang, D.M. Cunnold, L.P. Steele, P.B. Krummel, R.F. Weiss, S. O'Doherty, P.K. Salameh, H.J. Wang, C.M. Harth, J. Mühle, and L.W. Porter, 2008b, Renewed growth of atmospheric methane, *Geophysical Research Letters*, **35**, L22805, doi:10.1029/2008GL036037.

Riley, W.J., D.Y. Hsueh, J.T. Randerson, M.L. Fischer, J.G. Hatch, D.E. Pataki, W. Wang, and M.L. Goulden, 2008, Where do fossil fuel carbon dioxide emissions from California go? An analysis based on radiocarbon observations and an atmospheric transport model, *Journal of Geophysical Research*, **113**, G04002, doi:10.1029/2007JG000625.

Rivier, L., P. Ciais, D.A. Hauglustaine, P.S. Bakwin, P. Bousquet, P. Peylin, and A. Klonecki, 2006, Evaluation of SF_6, C_2Cl_4, and CO to approximate fossil fuel CO_2 in the northern hemisphere using a chemistry transport model, *Journal of Geophysical Research*, **111**, doi:10.1029/2005JD006725.

Robson, J.I., L.K. Gohar, M.D. Hurley, K.P. Shine, and T.J. Wallington, 2006, Revised IR spectrum, radiative efficiency and global warming potential of nitrogen trifluoride, *Geophysical Research Letters*, **33**, L10817, doi:10.1029/2006GL026210.

Rödenbeck, C., S. Houweling, M. Gloor, and M. Heimann, 2003, CO_2 flux history 1982-2001 inferred from atmospheric data using a global inversion of atmospheric transport, *Atmospheric Chemistry and Physics*, **3**, 2575-2659.

Rowland, F.S., S.C. Tyler, D.C. Montague, and Y. Makide, 1982, Dichlorodifluoromethane, CF_2Cl_2, in the Earth's atmosphere, *Geophysical Research Letters*, **9**, 481-484.

Roy, D.P., L. Boschetti, C.O. Justice, and J. Ju, 2008, The Collection 5 MODIS Burned Area Product: Global evaluation by comparison with the MODIS Active Fire Product, *Remote Sensing of Environment*, **112**, 3690-3707, doi:10.1016/j.rse.2008.05.013.

Rypdal, K., and W. Winiwarter, 2001, Uncertainties in greenhouse gas emissions inventories—Evaluation, comparability, and implications, *Environmental Science and Policy*, **4**, 107-116.

Saatchi, S.S., R.A. Houghton, R. Alvala, J.V. Soares, and Y. Yu, 2007, Distribution of aboveground live biomass in the Amazon basin, *Global Change Biology*, **13**, 816-837.

Sakamoto, T., P. Van Cao, N. Van Nguyen, A. Koter, and M. Yokozawa, 2009, Agro-ecological interpretation of rice cropping systems in flood-prone areas using MODIS imagery, *Photogrammetric Engineering and Remote Sensing*, **74**, 413-424.

Salas, W., S. Boles, C. Li, J.B. Yeluripati, X. Xiangming, S. Frolking, and P. Green, 2007, Mapping and modeling of greenhouse gas emissions from rice paddies with satellite radar observations and the DNDC biogeochemical model, *Aquatic Conservation: Marine and Freshwater Ecosystems*, **17**, 319-329.

Schuur, E.A.G., J.G. Vogel, K.G. Crummer, H. Lee, J.O. Sickman, and T.E. Osterkamp, 2009, The effect of permafrost thaw on old carbon release and net carbon exchange from tundra, *Nature*, **459**, 556-559.

Searchinger, T., R. Heimlich, R.A. Houghton, F.X. Dong, A. Elobeid, J. Fabiosa, S. Tokgoz, D. Hayes, and T.H. Yu, 2008, Use of U.S. croplands for biofuels increases greenhouse gases through emissions from land-use change, *Science*, **319**, 1238-1240.

Searchinger, T.D., S.P. Hamburg, J. Melillo, W. Chameides, P. Havlik, D.M. Kammen, G.E. Likens, R.N. Lubowski, M. Obersteiner, M. Oppenheimer, G.P. Robertson, W.H. Schlesinger, and G.D. Tilman, 2009, Fixing a critical climate accounting error, *Science* **326**, 527-528.

Serbin, G., C.S.T. Daughtry, E.R. Hunt Jr., J.B. Reeves III, and D.J. Brown, 2009, Effects of soil composition and mineralogy on remote sensing of crop residue cover, *Remote Sensing of Environment*, **113**, 224-238.

Simmonds, P.G., A.J. Manning, D.M. Cunnold, A. McCulloch, S. O'Doherty, R.G. Derwent, P.B. Krummel, P.J. Fraser, B. Dunse, L.W. Porter, R.H.J. Wang, B.R. Greally, B.R. Miller, P. Salameh, R.F. Weiss, and R.G. Prinn, 2006, Global trends, seasonal cycles, and European emissions of dichloromethane, trichloroethene, and tetrachloroethene from the AGAGE observations at Mace Head, Ireland, and Cape Grim, Tasmania, *Journal of Geophysical Research*, **111**, D18304, doi:10.1029/2006JD007082.

Skole, D., and C.J. Tucker, 1993, Tropical deforestation and habitat fragmentation in the Amazon: Satellite data from 1978-1988, *Science*, **260**, 1905-1910.

Smith, P., D. Martino, Z. Cai, D. Gwary, H. Janzen, P. Kumar, B. McCarl, S. Ogle, F. O'Mara, C. Rice, B. Scholes, and O. Sirotenko, 2007, Agriculture, in *Climate Change 2007: Mitigation of Climate Change*, Contribution of Working Group III to the Fourth Assessment Report of the Intergovernmental Panel on Climate Change, B. Metz, O.R. Davidson, P.R. Bosch, R. Dave, and L.A. Meyer, eds., Cambridge University Press, Cambridge, U.K., pp. 497-540.

Stajner, I., L.P. Riishøjgaard, and R.B. Rood, 2001, The GEOS ozone data assimilation system: Specification of error statistics, *Quarterly Journal of the Royal Meteorological Society*, **127**, 1069-1094.

Stavrakou, T., J.-F. Mueller, K.F. Boersma, I. De Smedt, and R.J. van der A, 2008, Assessing the distribution and growth rates of NO_x emission sources by inverting a 10-year record of NO_2 satellite columns, *Geophysical Research Letters*, **35**, L10801, doi:10.1029/ 2008GL033521.

Stinton, J.E., 2001, Accuracy and reliability of China's energy statistics, *China Economic Review*, **12**, 373-383.

Stohl, A., P. Seibert, J. Arduini, S. Eckhardt, P. Fraser, B.R. Greally, C. Lunder, M. Maione, J. Mühle, S. O'Doherty, R.G. Prinn, S. Reimann, T. Saito, N. Schmidbauer, P.G. Simmonds, M.K. Vollmer, R.F. Weiss, and Y. Yokouchi, 2009, An analytical inversion method for determining regional and global emissions of greenhouse gases: Sensitivity studies and application to halocarbons, *Atmospheric Chemistry and Physics*, **9**, 1597-1620.

Sullivan, D.G., J.N. Shaw, D. Rickman, P.L. Mask, and J.C. Luvall, 2005, Using remote sensing data to evaluate surface soil properties in Alabama ultisols, *Soil Science*, **170**, 954-968.

Sutton, M.A., E. Nemitz, J.W. Erisman, and 44 coauthors, 2007, Challenges in quantifying biosphere-atmosphere exchange of nitrogen species, *Environmental Pollution*, **150**, 125-139.

Swart, R., P. Bergamaschi, T. Pulles, and F. Raes, 2007, Are national greenhouse gas emissions reports scientifically valid? *Climate Policy*, **7**, 535-538.

Tangborn, A., I. Stajner, M. Buchwitz, I. Khlystova, S. Pawson, J. Burrows, R. Hudman, and P. Nedelec, 2009, Assimilation of SCIAMACHY total column CO observations: Global and regional analysis of data impact, *Journal of Geophysical Research*, **114**, D07307, doi:10.1029/2008JD010781.

Tans, P., I. Fung, and T. Takahashi, 1990, Observational constraints on the global atmospheric carbon dioxide budget, *Science*, **247**, 1431-1438.

Tansey, K., J.-M. Grégoire, P. Defourny, R. Leigh, J.-F. Pekel, E. van Bogaert, and E. Bartholomé, 2008, A new, global, multi-annual (2000-2007) burnt area product at 1 km resolution, *Geophysical Research Letters*, **35**, L011401, doi:10.1029/2007GL031567.

Thimsuwan, Y., A. Eiumnoh, K. Honda, and T. Tingsanchali, 2000, Estimation of methane emission from a deep-water rice field using Landsat TM and NOAA AVHRR: A case study of Bangkok Plain, *Imaging Science Journal*, **48**, 77-85.

Tilman, D., R. Socolow, J.A. Foley, J. Hill, E. Larson, L. Lynd, S. Pacala, J. Reilly, T. Searchinger, C. Somerville, and R. Williams, 2009, Beneficial biofuels—The food, energy, and environment trilemma, *Science*, **325**, 270-271.

Townshend, J.R., C. Huang, J.G. Masek, M.C. Hansen, S.N. Goward, C.J. Tucker, P. Davis, and S. Channan, 2008, An approach for developing Earth science data records of global forest cover change, *American Geophysical Union, Fall Meeting 2008*, abstract #IN54A-03.

Treuhaft, R.N., B.E. Law, and G.P. Asner, 2004, Forest attributes from radar interferometric structure and its fusion with optical remote sensing, *BioScience*, **54**, 561-572.

Turnbull, J., J.B. Miller, S.J. Lehman, P.P. Tans, R.J. Sparks, and J. Southon, 2006, Comparison of $^{14}CO_2$, CO, and SF_6 as tracers for recently added fossil fuel CO_2 in the atmosphere and implications for biological CO_2 exchange, *Geophysical Research Letters*, **33**, L01817, doi:10.1029/2005GL024213.

Turnbull, J.C., S.J. Lehman, J.B. Miller, R.J. Sparks, J.R. Southon, and P.P. Tans, 2007, A new high precision $^{14}CO_2$ time series for North American continental air, *Journal of Geophysical Research*, **112**, D11310, doi:10.1029/2006JD008184.

UNFCCC (United Nations Framework Convention on Climate Change), 2008, Greenhouse gas inventory data—Comparisons by category, available at <http://unfccc.int/di/Detailedby Category.do>.

United Nations, 1992, *United Nations Framework Convention on Climate Change*, FCCC/INFORMAL/84 GE.05-62220 (E) 200705, 24 pp., available at <http://unfccc.int/essential_background/convention/background/items/2853.php>.

United Nations, 1998, *Kyoto Protocol to the United Nations Framework Convention on Climate Change*, 20 pp., available at <http://unfccc.int/essential_background/kyoto_protocol/background/items/1351.php>.

United Nations, 2008a, *Report of the Australian Bureau of Statistics on Climate Change and Official Statistics*, Statistical Commission, United Nations Economic and Social Council, E/CN.3/2009/2, December 16, 33 pp.

United Nations, 2008b, *Towards International Recommendations for Energy Statistics*, Statistical Commission, United Nations Economic and Social Council, E/CN.3/2009/4, 15 pp.

United Nations, 2008c, *Oslo Group on Energy Statistics*, Statistical Commission, United Nations Economic and Social Council, E/CN.3/2009/5, 8 pp.

Uppala, S.M., P.W. Kållberg, A.J. Simmons, U. Andrae, V. DaCosta Bechtold, M. Fiorino, J.K. Gibson, J. Haseler, A. Hernandez, G.A. Kelly, X. Li, K. Onogi, S. Saarinen, N. Sokka, R.P. Allan, E. Andersson, K. Arpe, M.A. Balmaseda, A.C.M. Beljaars, L. Van De Berg, J. Bidlot, N. Bormann, S. Caires, F. Chevallier, A. Dethof, M. Dragosavac, M. Fisher, M. Fuentes, S. Hagemann, E. Hólm, B.J. Hoskins, L. Isaksen, P.A.E.M. Janssen, R. Jenne, A.P. McNally, J.-F. Mahfouf, J.-J. Morcrette, N.A. Rayner, R.W. Saunders, P. Simon, A. Sterl, K.E. Trenberth, A. Untch, D. Vasiljevic, P. Viterbo, and J. Woollen, 2005, The ERA-40 re-analysis, *Quarterly Journal of the Royal Meteorological Society*, **131**, 2961-3012.

Urbanski, S., C. Barford, S. Wofsy, C. Kucharik, E. Pyle, J. Budney, K. McKain, D. Fitzjarrald, M. Czikowsky, and J.W. Munger, 2007, Factors controlling CO_2 exchange on timescales from hourly to decadal at Harvard Forest, *Journal of Geophysical Research*, **112**, G02020, doi:10.1029/2006JG000293.

van der Gon, H.D., 1999, Changes in CH_4 emission from rice fields from 1960 to 1990s—2. The declining use of organic inputs in rice farming, *Global Biogeochemical Cycles*, **13**, 1053-1062.

van der Werf, G.R., J.T. Randerson, G.J. Collatz, L. Giglio, P.S. Kasibhatla, A.F. Arellano Jr., S.C. Olsen, and E.S. Kasischke, 2004, Continental-scale partitioning of fire emissions during the 1997 to 2001 El Niño/La Niña period, *Science*, **303**, 73-76.

van der Werf, G.R., J.T. Randerson, L. Giglio, G.J. Collatz, P.S. Kasibhatla, and A.F. Arellano Jr., 2006, Interannual variability in global biomass burning emissions from 1997 to 2004, *Atmospheric Chemistry and Physics*, **6**, 3423-3441.

van der Werf, G.R., J. Dempewolf, S.N. Trigg, J.T. Randerson, P.S. Kasibhatla, L. Giglio, D. Murdiyarso, W. Peters, D.C. Morton, G.J. Collatz, A.J. Dolman, and R.S. DeFries, 2008, Climate regulation of fire emissions and deforestation in equatorial Asia, *Proceedings of the National Academy of Sciences*, **105**, 20,350-20,355.

van der Werf, G.R., D.C. Morton, R.S. DeFries, J.G.J. Olivier, P.S. Kasibhatla, R.B. Jackson, G.J. Collatz, and J.T. Randerson, 2009a, CO_2 emissions from forest loss, Commentary, *Nature Geoscience*, **2**, 737-738.

van der Werf, G.R., D.C. Morton, R.S. DeFries, L. Giglio, J.T. Randerson, G.J. Collatz, and P.S. Kasibhatla, 2009b, Estimates of fire emissions from an active deforestation region in the southern Amazon based on satellite data and biogeochemical modeling, *Biogeosciences*, **6**, 235-249.

Velders, G.J.M., S. Madronich, C. Clerbaux, R. Derwent, M. Grutter, D. Hauglustaine, S. Incecik, M. Ko, J.-M. Libre, O.J. Nielsen, F. Stordal, and T. Zhu, 2005, Chemical and radiative effects of halocarbons and their replacement compounds, in *Safeguarding the Ozone Layer and the Global Climate System: Issues Related to Hydrofluorocarbons and Perfluorocarbons*, Special Report of the Intergovernmental Panel on Climate Change, B. Metz, L. Kuijpers, S. Solomon, S.O. Anderson, O. Davidson, J. Pons, D.

de Jager, T. Kestin, M. Manning, and L. Meyer, eds., Cambridge University Press, New York, pp. 133-180.

Vetter, M., G. Churkina, M. Jung, M. Reichstein, S. Zaehle, A. Bondeau, Y. Chen, P. Ciais, F. Feser, A. Freibauer, R. Geyer, C. Jones, D. Papale, J. Tenhunen, E. Tomelleri, K. Trusilova, N. Viovy, and M. Heimann, 2008, Analyzing the causes and spatial pattern of the European 2003 carbon flux anomaly using seven models, *Biogeosciences*, **5**, 561-583.

Vogelmann, J.E., S.M. Howard, L. Yang, C.R. Larson, B.K. Wylie, and N. Van Driel, 2001, Completion of the 1990s National Land Cover Dataset for the coterminous United States from Landsat Thematic Mapper data and ancillary data sources, *Photogrammetric Engineering and Remote Sensing*, **67**, 650-662.

Vuichard, N., P. Ciais, N. Viovy, P. Calanca, and J.F. Soussana, 2007, Estimating the greenhouse gas fluxes of European grasslands with a process-based model: 2. Simulations at the continental level, *Biogeochemical Cycles*, **21**, GB1005, doi:10.1029/2005GB002612.

Walter, K.M., S.A. Zimov, J.P. Chanton, D. Verbyla, and F.S. Chapin III, 2006, Methane bubbling from Siberian thaw lakes as a positive feedback to climate warming, *Nature*, **443**, 71-75.

Wang, J.-W., A.S. Denning, L. Lu, I.T. Baker, K.D. Corbin, and K.J. Davis, 2007, Observations and simulations of synoptic, regional, and local variations in atmospheric CO_2, *Journal of Geophysical Research*, **112**, D04108, doi:10.1029/2006JD007410.

Wang, W., I. Kazuhito Ichii, H. Hashimoto, A.R. Michaelis, P.E. Thornton, B.E. Law, and R.R. Nemani, 2009, A hierarchical analysis of terrestrial ecosystem model Biome-BGC: Equilibrium analysis and model calibration, *Ecological Modeling*, **220**, 2009-2023.

Wassmann, R., H.U. Neue, M.C.R. Alberto, R.S. Lantin, C. Bueno, D. Llenaresas, J.R.M. Arah, H. Papen, W. Seiler, and H. Rennenberg, 1996, Fluxes and pools of methane in wetland rice soils with varying organic inputs, *Environmental Monitoring and Assessment*, **42**, 163-173.

Watson, A.J., U. Schuster, D.C.E. Bakker, N.R. Bates, A. Corbière, M. González-Dávila, T. Friedrich, J. Hauck, C. Heinze, T. Johannessen, A. Körtzinger, N. Metzl, J. Olafsson, A. Olsen, A. Oschlies, X.A. Padin, B. Pfeil, J.M. Santana-Casiano, T. Steinhoff, M. Telszewski, A.F. Rios, D.W.R. Wallace, and R. Wanninkhof, 2009, Tracking the variable North Atlantic sink for atmospheric CO_2, *Science*, **326**, 1391-1393.

Weiss, R.F., J. Mühle, P.K. Salameh, and C.M. Harth, 2008, Nitrogen trifluoride in the global atmosphere, *Geophysical Research Letters*, **35**, L20821, doi:10.1029/2008GL035913.

West, T.O., C.C. Brandt, V.S. Wilson, C.M. Hellwinckel, D.D. Tyler, G. Marland, D.G. De La Torre Ugarte, J.A. Larson, and R.G. Nelson, 2008, Estimating regional changes in soil carbon with high spatial resolution, *Soil Science Society of America Journal*, **72**, 285-294.

Wiedinmyer, C., B. Quayle, C. Geron, A. Belote, D. McKenzie, X. Zhang, S. O'Neill, and K.K. Wynne, 2006, Estimating emissions from fires in North America for air quality modeling, *Atmospheric Environment*, **40**, 3419-3432.

Wild, O., M.J. Prather, and H. Akimoto, 2001, Indirect long-term global cooling from NO_x emissions, *Geophysical Research Letters*, **28**, 1719-1722.

Wooster, M.J., G. Roberts, G.L.W. Perry, and Y.J. Kaufman, 2005, Retrieval of biomass combustion rates and totals from fire radiative power observations: FRP derivation and calibration relationships between biomass consumption and fire radiative energy release, *Journal of Geophysical Research*, **110**, D24311, doi:10.1029/2005JD006318.

Wunch, D., P.O. Wennberg, G.C. Toon, G. Keppel-Aleks, and Y.G. Yavin, 2009, Emissions of greenhouse gases from a North American megacity, *Geophysical Research Letters*, **36**, L15810, doi:10.1029/2009GL039825.

Yadav, V., and G. Malanson, 2007, Progress in soil organic matter research: Litter decomposition, modeling, monitoring and sequestration, *Progress in Physical Geography*, **31**, 131-154.

Yao, H., Z. Wen, Z. Xunhua, H. Shenghui, and Y. Yongqiang, 2006, Estimates of methane emissions from Chinese rice paddies by linking a model to GIS database, *Acta Ecological Sinica*, **26**, 980-988.

Zhang, Q., D.G. Streets, K. He, Y. Wang, A. Richter, J.P. Burrows, I. Uno, C.J. Jang, D. Chen, Z. Yao, and Y. Lei, 2007, NO_x emissions trends for China, 1995-2004: The view from the ground and the view from space, *Journal of Geophysical Research*, **112**, D22306, doi:10.1029/2007JD008684.

Zhao, C., A.E. Andrews, L. Bianco, J. Eluszkiewicz, A. Hirsch, C. MacDonald, T. Nehrkorn, and M.L. Fischer, 2009, Atmospheric inverse estimates of methane emissions from central California, *Journal of Geophysical Research*, **114**, D16302, doi:10.1029/2008JD011671.

Zimov, S.A., S.P. Davydov, G.M. Zimova, A.I. Davydova, E.A.G. Schuur, K. Dutta, and F.S. Chapin, 2006, Permafrost carbon: Stock and decomposability of a globally significant carbon pool, *Geophysical Research Letters*, **33**, L20502, doi:10.1029/2006GL027484.

Appendixes

Appendix A

UNFCCC Inventories of Industrial Processes and Waste

INDUSTRIAL PROCESSES AND PRODUCT USE

The industrial processes and product use (IPPU) sector covers the greenhouse gas emissions resulting from various industrial activities that produce emissions not directly the result of energy consumed during the process and the use of man-made greenhouse gases in products (IPCC, 2006). Examples include the release of CO_2 as a by-product of cement production and the use of fossil fuel (primarily natural gas) as a feedstock in ammonia production. The IPPU sector accounts for about 7 percent of total greenhouse gas emissions from Annex I countries (UNFCCC, 2008) and about 6 percent of total greenhouse gas emissions for non-Annex I countries (UNFCCC, 2005).

Carbon Dioxide

Carbon dioxide (CO_2) is the most important greenhouse gas emitted by the IPPU sector, comprising about 69 percent of total emissions (in terms of CO_2 equivalents) from the sector for Annex I countries (UNFCCC, 2005). The main sources of CO_2 in this sector are the production of cement, lime, glass, ammonia, iron, steel, and aluminum. The calcination of limestone produces lime, which may then be combined with silica compounds to produce clinker (an ingredient of cement). Both processes result in CO_2 emissions. Glass production emits CO_2 during the melting and fusion of limestone, dolomite, and soda ash. The principal source of CO_2 emissions from ammonia production is the steam reforming of natural gas (methane, CH_4) to produce hydrogen (H_2). Iron and steel production yields CO_2 emissions through the use of metallurgical coke to convert iron ore to pig iron in a blast furnace. Similarly, CO_2 is emitted during the smelting process from the use of carbon to reduce alumina to aluminum.

The CO_2 emissions from mineral, chemical, and metal production can be estimated simply by applying appropriate emission factors to national-level production data. The major source of uncertainty in emissions from the mineral industry is typically the activity data (IPCC, 2006; EPA, 2008) because the chemistry of the processes involved is known. For cement production, CO_2 emissions should ideally be estimated using national-level data on clinker production, the lime content of the clinker, and the fraction of lime from limestone. However, national statistics on cement and/or clinker production may not be complete for countries in which a substantial part of production comes from numerous small kilns, for which data are difficult to obtain. If clinker production data are not available, they are inferred from information on the quantities of cement produced (correcting for imports and exports) and the types and clinker fraction of the cement. For lime and glass production, CO_2 emissions can be estimated using national-level data on the types and quantities of lime or glass produced (or, less preferably, total lime or glass production figures) and default emission factors. The key source of uncertainty for lime production is incomplete data; reported lime production statistics often omit nonmarketed lime production, potentially resulting in order-of-magnitude underestimates. For glass production, activity data uncertainties are magnified where glass production is measured in a variety of units.

In the chemical and metal industries, reliable production data are available for most countries, so the emission factors present the greatest source of uncertainty, particularly for iron and steel production. For ammonia production, CO_2 emissions can be estimated using national-level data on ammonia production (or, less preferably, ammonia production capacity) and default values for the quantity of fuel (typically natural gas) required as feedstock per unit of output, the carbon content of the fuel, and the carbon oxidation factor. Any CO_2 recovered for purposes of urea production is also accounted for. For iron and steel production, the CO_2 emissions are estimated by applying the appropriate emission factors to national statistics on the amount of steel produced by each method and the total amount of pig iron produced that is not processed into steel. Similarly, the estimation of CO_2 emissions in aluminum production requires national-level production data by process type (i.e., Søderberg or Prebake) to which the appropriate default emission factor can then be applied.

Hydrofluorocarbons

Hydrofluorocarbons (HFCs) comprise about 18 percent of total emissions (in terms of CO_2 equivalents) from the IPPU sector for Annex I countries.[1] The use of HFCs as substitutes for ozone-depleting substances in a variety of industrial applications is by far the largest source of HFC emissions, accounting for about 86 percent of total emissions from the sector, and their usage is growing rapidly. A smaller, but significant source of HFC emissions is the generation of trifluoromethane (HFC-23) as a by-product during the production of chlorodifluoromethane (HCFC-22).

Actual emissions of HFCs are estimated using either an emission-factor or a mass-balance approach (IPCC, 2006). Both methods can use activity data collected at either the application level (e.g., refrigeration) or the subapplication level (e.g., equipment or product type); the latter is expected to yield higher-accuracy estimates. For the emission-factor approach, HFC emissions are calculated by determining the net consumption of a chemical in a specific application or subapplication (production plus imports minus exports minus destruction of the chemical) and then applying an emission factor(s) to the net consumption. For the mass-balance approach, emissions are estimated as the sum of the sales of a chemical and, for equipment containing this chemical, the total charge of retired equipment minus the total charge of new equipment. The major source of uncertainty in national estimates of HFC emissions is the lack of activity data on chemical production or sales in countries where suppliers treat the information as confidential. This barrier to the production of reliable national estimates is being reduced with the development of regional and global databases of ozone-depleting substances. For example, databases that track the phase-out of ozone-depleting substances are directly relevant for estimating the phase-in of HFC substitutes (IPCC, 2006).

Emissions of HFC-23 can be calculated by applying a default emission factor to the quantity of HCFC-22 produced. Given the known variability in emissions from different HCFC-22 manufacturing facilities, the uncertainty in the emission factor far outweighs the uncertainty in the activity data (IPCC, 2006).

Nitrous Oxide

Emissions of nitrous oxide (N_2O) from nitric acid and adipic acid production comprise about 7 percent of total emissions from the IPPU sector in Annex I countries.[2] Nitric acid production emits N_2O as a by-product during the catalytic oxidation of ammonia, and adipic acid production (most of which takes place in a few plants in the United States and Europe) generates N_2O as a by-product during a process involving the oxidation of nitric acid. Emissions of N_2O from both sources can be estimated by multiplying production by a default emission factor. For nitric acid production, the major source of uncertainty in N_2O emissions is the activity data. Nitric acid production is often underestimated because nitric acid is formed as part of a larger production process and is never sold on the market. For adipic acid production, neither the default emission factor nor the activity data are significant sources of uncertainty (IPCC, 2006). The default emission factor

---

[1] See <http://unfccc.int/di/DetailedByGas.do>.

[2] See <http://unfccc.int/di/DetailedByGas.do>.

is derived from a well-understood chemical reaction (i.e., nitric acid oxidation), and only a small number of adipic acid plants exist.

WASTE

The waste sector is not a significant source of greenhouse gas emissions, accounting for only about 3 percent of the total from Annex I countries (UNFCCC, 2008) and about 4 percent of the total from non-Annex I countries (UNFCCC, 2005). This sector covers the greenhouse gas emissions from solid waste disposal, biological treatment of solid waste, burning of waste, and wastewater treatment and discharge (IPCC, 2006). Waste sector reporting includes neither the greenhouse gas emissions resulting from the use of waste material as fuel nor the CO_2 emissions resulting from the decomposition or burning of organic biomass. These emissions are accounted for under the energy sector and the agriculture, forestry, and other land-use sector, respectively.

Methane

The primary greenhouse gas emitted from the waste sector is CH_4, which accounts for about 90 percent of the total (in terms of waste sector CO_2 equivalents) in Annex I countries.[3] The degradation of organic material under anaerobic conditions at solid waste disposal sites (SWDS) is the principal source of CH_4 emissions. The potential of SWDS to generate CH_4 depends on the degradable organic carbon (DOC) content of the waste, which is a function of the amount and composition of the waste disposed, and on the waste management practices. Methane emissions from SWDS are calculated using the First Order Decay method, which assumes that the rate of CH_4 production is directly proportional to the amount of DOC remaining in the waste. The quantity of CH_4 that is oxidized in the landfill's top layer and/or is recovered and combusted is then subtracted from the calculated emissions value.

The key source of uncertainty in estimates of CH_4 from SWDS is the activity data relating to the quantities and composition of the waste disposed (several decades of historical data are required), although some of the emission factors can also be highly uncertain. For many countries, data on waste amounts and composition (particularly historical data) are not available and default activity data must be used. The major uncertainties in the emission factors include the DOC values assigned to different waste types (e.g., municipal) and materials (e.g., paper, food), the fraction of DOC that is ultimately degraded and released from SWDS, and the half-life of the DOC, which is difficult to measure in real solid waste disposal sites. Also highly uncertain are the emission factors used to determine the fraction of CH_4 that is oxidized in the landfill's top layer, which depends on whether the SWDS is managed or unmanaged and also varies considerably with conditions at the site.

The other significant source of CH_4 emissions within the waste sector is the anaerobic treatment or disposal of wastewater. The CH_4 emitted from wastewater handling depends on the amount of degradable organic material, measured by biological oxygen demand in domestic wastewater and chemical oxygen demand in industrial wastewater. The Intergovernmental Panel on Climate Change (IPCC) provides a means of estimating the quantity of domestic wastewater generated as well as default values for biological oxygen demand for selected regions and countries. Similarly, the IPCC provides default values for quantities of industrial wastewater generated and the chemical oxygen demand for various industry types. Reliable estimates of the quantity of CH_4 released from wastewater discharge are particularly difficult to obtain for developing countries due to uncertainties in the fraction of domestic wastewater that is removed by sewers (as opposed to being treated in latrines), the fraction of sewers that are open, and the degree to which these open sewers are anaerobic (IPCC, 2006).

Carbon Dioxide

Carbon dioxide is a relatively minor source of greenhouse gas emissions from the waste sector, accounting for about 4 percent of total emissions (in terms of CO_2 equivalents) from the sector for Annex I Parties.[4] The predominant source of these emissions,

---

[3] See <http://unfccc.int/di/DetailedByGas.do>.

[4] See <http://unfccc.int/di/DetailedByGas.do>.

comprising about 97 percent of total CO_2 emissions from this sector, is the incineration and open burning of waste containing fossil carbon (e.g., plastics, certain textiles). The practice of waste incineration is currently more common in developed countries, while open burning of waste occurs predominantly in the developing world. However, the basic approach recommended by the IPCC for estimating CO_2 emissions from these two sources is the same: the quantity of waste incinerated and/or open-burned is multiplied by default values for the dry matter content, total carbon content, fossil carbon fraction, and oxidation factor for the waste (IPCC, 2006). The major source of uncertainty is the estimation of the fossil carbon fraction of the waste, which is directly related to uncertainties regarding waste composition. Where country-specific data regarding quantities of waste incinerated and/or open-burned are not available, large uncertainties are also associated with the waste amounts determined from the IPCC default values for waste generation and management.

Nitrous Oxide

Nitrous oxide emissions comprise about 6 percent of total emissions (in terms of CO_2 equivalents) from the waste sector for Annex I countries.[5] The major source, comprising about 82 percent of total N_2O emissions from the sector, is wastewater handling. N_2O is emitted from the degradation of nitrogen components in the wastewater (e.g., urea, nitrate, protein). Although both wastewater treatment plants and the discharge of effluent into aquatic environments are sources of N_2O emissions, the latter is typically a far more significant source. Emissions of N_2O from wastewater effluent discharged to aquatic environments are determined using national statistics on population and annual per capita protein consumption to estimate the total amount of nitrogen discharged in wastewater effluent, and a default emission factor for the N_2O emitted per unit of wastewater effluent nitrogen content (IPCC, 2006). Large uncertainties are associated with estimates of N_2O emissions from wastewater handling, and the major source of uncertainty is the default emission factor for N_2O from the effluent.

REFERENCES

EPA (Environmental Protection Agency), 2008, *Inventory of U.S. Greenhouse Gas Emissions and Sinks: 1990-2006*, EPA 430-R-08-005, Office of Atmospheric Programs, Washington, D.C., available at <http://www.epa.gov/climatechange/emissions/usgginventory.html>.

IPCC (Intergovernmental Panel on Climate Change), 2006, *2006 IPCC Guidelines for National Greenhouse Gas Inventories*, H.S. Eggleston, L. Buendia, K. Miwa, T. Ngara, and K. Tanabe, eds., prepared by the National Greenhouse Gas Inventories Programme, Institute for Global Environmental Strategies, Hayama, Kanagawa, Japan, 5 volumes.

UNFCCC (United Nations Framework Convention on Climate Change), 2005, Sixth compilation and synthesis of initial national communications from Parties not included in Annex I to the Convention, prepared by the UNFCCC Secretariat, October 2005, available at <http://unfccc.int/ghg_data/ghg_data_unfccc/items/4146.php>.

UNFCCC, 2008, Report on national greenhouse gas inventory data from Parties included in Annex I to the Convention for the period 1990-2006, prepared by the UNFCCC Secretariat, November 2008, available at <http://unfccc.int/ghg_data/ghg_data_unfccc/items/4146.php>.

---

[5] See <http://unfccc.int/di/DetailedByGas.do>.

Appendix B

Estimates of Signals Created in the Atmosphere by Emissions

To determine how well national emissions can be quantified from atmospheric measurements, it is necessary to first estimate the mole fraction signals produced over a country by its emissions, then compare the result to technical capabilities. The average emissions are defined as the total national emissions divided by the area, and they are expressed in units of moles per square meter per second (mol m^{-2} s^{-1}). An air mass with a certain thickness, traveling over a region, collects emissions during its transit. The transit time is estimated as the fetch divided by the wind speed. The fetch is taken as the square root of the area, and a wind speed of 5 m s^{-1} is assumed. When trace gas emissions are mixed into the air mass, its mole fraction changes (denoted by Δ); the change is smaller when the trace gas is diluted into a larger air mass. The latter is expressed as moles per square meter in the column of air into which the trace is being mixed (equation 1).

$$\frac{emissions\,(mol\,s^{-1})}{area} \times \frac{fetch\,(\sqrt{area})}{wind\,speed} \times \frac{1}{mol\,m^{-2}} = \Delta(dry\,mole\,fraction) \quad (1)$$

At sea level the total atmospheric column contains ~356,000 mol m^{-2} of air. The lowest 1 km, a typical height of the atmospheric boundary layer, contains ~40,000 mol m^{-2} of air at sea level, or ~11 percent of the total atmosphere. In Table B.1 the total column and the boundary layer masses assume the surface to be sea level for each country. At an average wind speed of 5 m s^{-1}, air moves 430 km in 24 hours, so it takes 4.6 days to traverse a fetch of 2,000 km. Air usually remains in the boundary layer 3-5 days. For longer fetches it becomes increasingly unlikely that all of the emissions remain confined to the boundary layer, and the mole fraction change is then overestimated for those cases.

Table B.1 compares national carbon dioxide (CO_2) emissions and the estimated atmospheric signal for the 20 largest CO_2-emitting nations, which represent more than 80 percent of estimated global emissions. The numbers in the last two columns are typical, but will vary greatly in practice because they are inversely proportional to wind speed. How do the estimates in Table B.1 compare with observations? Based on a very limited number of ^{14}C measurements of CO_2 in small aircraft off the coast of Cape May and New Hampshire, the average component of CO_2 from recent fossil-fuel combustion is ~3 parts per million (ppm) in the boundary layer.

It is striking to note in Table B.1 how small, relative to background, the mole fraction signals are when averaged over an entire country, and also that the signals for Japan, Germany, and Korea are comparable to those for the two largest emitters: the United States and China. That is due to the higher emissions intensity (per square meter) in the three smaller countries.

Table B.2 presents data from the same 20 countries, for methane (CH_4), nitrous oxide (N_2O), and sulfur hexafluoride (SF_6), estimated in the same way as in Table B.1. When comparing these estimates with U.S. observations from the National Oceanic and Atmospheric Administration network, the estimated

TABLE B.1 National CO_2 Emissions and Estimated Atmospheric Signal

Country	Area (1,000 km^2)	Fetch (km)	CO_2 (Mton yr^{-1})	Population (millions)	Average Emissions (μmol m^{-2} s^{-1})	Total Column (ppm)	Boundary Layer 1 km (ppm)
United States	9,827	3,134	5,892	307.0	0.432	0.76	6.8
China	9,571	3,093	5,577	1,329.0	0.420	0.73	6.5
Russia	17,075	4,132	1,568	142.0	0.066	0.15	1.4
Japan	378	614	1,249	128.0	2.379	0.82	7.3
India	3,166	1,779	1,222	1169.0	0.278	0.28	2.5
Germany	357	597	829	82.6	1.672	0.56	5.0
Canada	9,985	3,159	559	32.8	0.040	0.07	0.6
UK	244	493	539	60.8	1.591	0.44	3.9
Italy	301	548	477	58.9	1.141	0.35	3.1
South Korea	99	314	475	48.2	3.455	0.61	5.4
Iran	1,648	1,283	447	71.2	0.195	0.14	1.3
Mexico	1,964	1,401	411	106.5	0.151	0.12	1.1
France	544	737	399	61.6	0.528	0.22	1.9
Australia	7,682	2,771	382	20.7	0.036	0.06	0.5
Indonesia	1,905	1,380	367	232.0	0.139	0.11	1.0
Spain	506	711	367	44.3	0.522	0.21	1.9
Brazil	8,547	2,923	352	192.0	0.030	0.05	0.4
South Africa	1,219	1,104	337	48.6	0.199	0.12	1.1
Saudi Arabia	2,240	1,496	333	24.8	0.107	0.09	0.8
Ukraine	604	777	303	46.2	0.361	0.16	1.4

NOTES: Mton CO_2 = million metric ton of CO_2. The totals include electricity generation, heating, transportation, manufacturing and construction, and other fuel combustion, but not bunker fuels, land-use change, or waste.
SOURCES: National totals for 2005 from the World Resources Institute. Area and population are from the CIA Fact book.

TABLE B.2 National Emissions of CH_4, N_2O, and SF_6 and Estimated Atmospheric Signals

Country	CH_4				N_2O				SF_6			
	Emissions (Mton yr^{-1})	Average (nmol m^{-2} s^{-1})	Total Column (ppb)	Boundary Layer 1 km (ppb)	Emissions (Mton yr^{-1})	Average (pmol m^{-2} s^{-1})	Total Column (ppb)	Boundary Layer 1 km (ppb)	Emissions (ton yr^{-1})	Average (fmol m^{-2} s^{-1})	Total Column (ppt)	Boundary Layer 1 km (ppt)
United States	20.84	4.20	7.4	65	1.261	92	0.163	1.4	842	18.60	0.0327	0.29
China	34.13	7.06	12.3	109	2.296	172	0.300	2.7	364	8.25	0.0143	0.13
Russia	12.58	1.46	3.4	30	0.193	8	0.019	0.2	175	2.22	0.0052	0.05
Japan	0.84	4.40	1.5	13	0.119	226	0.078	0.7	57	32.73	0.0113	0.10
India	21.91	13.70	13.7	121	0.239	54	0.054	0.5	92	6.31	0.0063	0.06
Germany	2.74	15.20	5.1	45	0.213	429	0.144	1.3	35	21.28	0.0071	0.06
Canada	4.08	0.81	1.4	12	0.194	13	0.025	0.2	110	2.39	0.0042	0.04
UK	1.85	15.01	4.2	37	0.153	451	0.125	1.1	22	19.57	0.0054	0.05
Italy	1.38	9.08	2.8	24	0.150	358	0.111	1.0	18	12.98	0.0040	0.04
South Korea	1.34	26.80	4.7	42	0.057	414	0.073	0.7	96	210.45	0.0372	0.33
Iran	3.83	4.60	3.3	29	0.077	33	0.024	0.2	22	2.90	0.0021	0.02
Mexico	7.39	7.45	5.9	52	0.090	32	0.026	0.2	35	3.87	0.0030	0.03
France	2.44	8.88	3.7	32	0.268	354	0.147	1.3	18	7.18	0.0030	0.03
Australia	5.16	1.33	2.1	18	0.107	10	0.016	0.1	31	0.88	0.0014	0.01
Indonesia	7.32	7.61	5.9	52	0.142	53	0.042	0.4	18	2.05	0.0016	0.01
Spain	1.46	5.71	2.3	20	0.097	138	0.055	0.5	9	3.86	0.0015	0.01
Brazil	15.56	3.61	5.9	52	0.888	74	0.123	1.1	75	1.90	0.0031	0.03
South Africa	2.21	3.59	2.2	19	0.078	16	0.029	0.3	35	6.23	0.0039	0.03
Saudi Arabia	1.11	0.98	0.8	7	0.038	12	0.010	0.1	22	2.13	0.0018	0.02
Ukraine	6.14	20.13	8.8	78	0.091	108	0.047	0.4	31	11.14	0.0049	0.04

NOTES: nmol = nanomol, 10^{-9} mol; pmol = picomol, 10^{-12} mol; fmol = femtomol, 10^{-15} mol.
SOURCE: National emissions for 2005 from the World Resources Institute.

enhancement of greenhouse gas mole fractions is not far from, but on the high side of, observed values for the lowest 1 km, consistent with the fact that over long fetches the emissions tend not to remain confined to the boundary layer. For CH_4, the observed annual average enhancement in the boundary layer of the United States, relative to background values, is 20-60 parts per billion (ppb), with large standard deviation of midday means of 20-30 ppb. For N_2O, these numbers are 0.2-0.4 ppb, and daily standard deviation of 0.3-0.7 ppb; for SF_6, the average enhancement is 0.05-0.2 parts per trillion (ppt), with daily variability of 0.07 to 0.13 ppt. For N_2O, the highest observed enhancement is 0.7 ppb over Iowa, which is indicative of intense regional emissions. The World Meteorological Organization-recommended accuracies for in situ measurements are 2 ppb for CH_4, 0.1 ppb for N_2O, and 0.02 ppt for SF_6.

Table B.3 explores expected enhancements of the CO_2 mole fraction over metropolitan areas. The signal expected to be produced over a single large city relative to its surroundings is comparable to, and in many cases larger than, the average produced by an entire country. Note that the observed average fossil-fuel CO_2 enhancement at the surface in Los Angeles based on ^{14}C measurements of plants (Figure 4.5) is about five times larger over part of the basin than the number estimated in Table B.3.

The latter is derived with a standard assumption of a steady 5 m s^{-1} average wind vector, which would imply that the residence time of air over the metropolitan area would be ~4 hours. Because Los Angeles is surrounded by mountains on three sides, the residence time over the city is much longer.

SIGNAL FROM A 1 GW(E) COAL FIRED POWER PLANT

One gigawatt (GW) corresponds to 8.76 10^9 kWh yr^{-1}. Applying the U.S. average 25 mol C kWh^{-1}, the plant would produce 2.19 10^{11} mol yr^{-1}, or 6,900 mol s^{-1}. If the perpendicular distance across the plume is 1.7 km at some distance downwind, and the wind speed is 5 m s^{-1}, then 1 second of CO_2 emissions is diluted into 1,700 × 5 m^2 s^{-1} × 3.56 10^5 mol of air per square meter (full atmospheric column). The number 1.7 km is chosen to correspond to the 3 km^2 footprint of a single Orbiting Carbon Observatory (OCO) sounding. The CO_2 increase in the total column is then 2.3 ppm.

SIGNAL FROM A GEOLOGICAL SEQUESTRATION LEAK

Significant quantities of CO_2 may one day be captured at large point sources in the utility and industrial sectors and injected into storage sites in the Earth, rather than released to the atmosphere. If this occurs, it will be important to monitor both the quantity of CO_2 that is injected into storage sites and the amount of any CO_2 that leaks from these sites (IPCC, 2005). Leaks from geological sequestration are relatively easy to detect. The emissions are at the surface and are not buoyant like those from a power plant. At night they

TABLE B.3 Expected CO_2 Signals for Selected Metropolitan Areas

City	Area (km^2)[a]	Emissions (Mton CO_2 yr^{-1})	Emissions (μmol m^{-2} s^{-1})	Total Column (ppm)	Boundary Layer 1 km (ppm)
Los Angeles	3,700	73.2	14.2	0.49	4.3
Chicago	2,800	79.1	20.3	0.60	5.4
Houston	3,300	101.8	22.2	0.72	6.4
Indianapolis	900	20.1	16.1	0.27	2.4
Tokyo	1,700	64	27	0.63	5.6
Seoul	600	43	52	0.71	6.3
Beijing	800	74	67	1.1	9.4
Shanghai	700	112	116	1.8	15

NOTES: Mton CO_2 is million metric tons of CO_2.

[a]Area represents the contiguous area of intense and activity and was estimated using Google maps in "satellite" mode, which shows built up areas by color and road density.

SOURCES: Emissions in 1998 for four east Asian cities from Dhakal et al. (2003). U.S. estimates are from the VULCAN emissions inventory for 2002 (<www.purdue.edu/eas/carbon/vulcan>).

will tend to stay at the surface. Assume that the CO_2 from a 1 GW(e) power plant is captured and injected into a well. If 0.1 percent escapes during the injection, there would be a point source of 6.9 mol s^{-1} of CO_2. At a point 1 km downstream, with a wind speed of 5 m s^{-1} and a plume width of 100 m and plume height of 100 m, the CO_2 enhancement would be 3.3 ppm, causing a ^{14}C depletion of 0.89 percent. A second scenario is escape from the geological formation into which the CO_2 has been stored. If 0.2 percent of the sequestered CO_2 escapes per year (residence time 500 years), the leak would increase over time as more CO_2 is pumped into the formation. After 1 year the leak rate would be 14 mol s^{-1}, after 2 years 28 mol s^{-1}, and so on. In addition, when the escaping CO_2 goes through soil, the CO_2 mole fraction in soil air would become very high, eventually depriving the vegetation of sufficient oxygen in the root zone and leading to plant death, which should be easily detectable.

REFERENCES

Dhakal, S., S. Kaneko, and H. Imura, 2003, CO_2 emissions from energy use in East-Asian mega-cities, in *Proceedings of the International Workshop on Policy Integration Towards Sustainable Urban Use for Cities in Asia*, February 4-5, East-West Center, Honolulu, Hawaii, available at <http://enviroscope.iges.or.jp/contents/6/index.html>.

IPCC (Intergovernmental Panel on Climate Change), 2005, *Carbon Dioxide Capture and Storage*, IPCC Special Report, prepared by Working Group III of the Intergovernmental Panel on Climate Change, B. Metz, O. Davidson, H.C. de Coninck, M. Loos, and L.A. Meyer, eds., Cambridge University Press, New York, 442 pp.

Appendix C

Current Sources of Atmospheric and Oceanic Greenhouse Gas Data

ATMOSPHERIC DATA

The atmospheric monitoring sites of Mauna Loa and the South Pole, established during the 1957-1958 International Geophysical Year by C.D. Keeling, have been expanded to both remote and near continental sites (e.g., the ALE/GAGE [Atmospheric Lifetime Experiment-Global Atmospheric Gases Experiment] network, Prinn et al., 1983; the NOAA ESRL [National Oceanic and Atmospheric Administration Earth System Research Laboratory] network, Conway et al., 1994) and to include many other trace gases. The current global greenhouse gas measurement network is an international effort involving about 50 countries. It is coordinated by the World Meteorological Organization's (WMO's) Global Atmosphere Watch (GAW) Programme. The NOAA ESRL network, shown in Figure 4.2, is the largest contributing network to GAW. The WMO plays a crucial role in the international monitoring endeavor (1) by promulgating a common calibration scale for each species and quantitative goals for the comparability of measurements by participating laboratories; (2) by promoting comparison programs, measurement system audits, quality assurance and quality control guidelines, and submission of data to the World Data Center for Greenhouse Gases; and (3) by supporting capacity building.

Carbon dioxide (CO_2) data are also available from three satellites—SCIAMACHY (Scanning Imaging Absorption Spectrometer for Atmospheric Chartography), AIRS (Atmospheric Infrared Sounder), and IASI (Infrared Atmospheric Sounding Interferometer)—and soon will be available from GOSAT (Greenhouse gases Observing Satellite), the first satellite designed for measuring CO_2. A number of studies have used available satellite data to estimate atmospheric CO_2 (e.g., Crevoisier et al., 2004, 2009; Buchwitz et al., 2005; Chahine et al., 2005, 2008; Maddy et al., 2008; Schneising et al., 2008; Strow and Hannon, 2008). Table C.1 contains information about the satellites as well as the Orbiting Carbon Observatory (OCO), which failed at launch on February 24, 2009. Unlike AIRS and IASA, OCO and GOSAT had a calibration system in place and a weighting function in the lower troposphere where the signal from surface emissions is strongest. OCO's spatial resolution was more than an order of magnitude higher than any other satellite's (instantaneous field of view <3 km^2), and its signal-to-noise ratio was three times that of GOSAT.

OCEAN DATA

Carbon Dioxide

The accumulation of anthropogenic CO_2 in the ocean resulting from atmospheric uptake can be readily observed. The inventory today is more than 500 billion tons and is increasing at a rate of >1 million tons per hour. Unlike the atmosphere and land surface, the oceanic CO_2 signal is not amenable to satellite observation; seawater is a conducting medium and is impervious to electromagnetic radiation. Instead, CO_2 is measured during large-scale observing expeditions at approximately decadal intervals. A few time-series stations are also maintained at locations where the changes

TABLE C.1 Specifications of Spaceborne Instruments Capable of Measuring CO_2

Specification	OCO[a]	GOSAT[b]	SCIAMACHY[c]	AIRS[d]	IASI[e]
Tropospheric gases measured	CO_2, O_2	CO_2, CH_4, O_2, O_3, H_2O	O_3, O_4, N_2O, NO_2, CH_4, CO, CO_2, H_2O, SO_2, HCHO	CO_2, CH_4, O_3, CO, H_2O, SO_2	CO_2, CH_4, O_3, CO, H_2O, SO_2, N_2O
CO_2 sensitivity	Total column including near surface	Total column including near surface	Total column including near surface	Midtroposphere	Midtroposphere
Horizontal resolution (km)[f]	1.29 × 2.25/5.2	FTS: 10.5/80-790	30 × 60/960	15/1,650	12/2,200
CO_2 uncertainty (ppm)[g]	1-2	4	14	1.5	2
Instruments	3-channel grating spectrometer	CAI, SWIR/TIR Fourier transform spectrometer	8-channel grating spectrometer	Grating spectrometer	Fourier transform spectrometer
Viewing modes	Nadir, glint, target	Nadir, glint, target	Limb, nadir	Nadir	Nadir
Samples per day	500,000	18,700	8,600	2,916,000	1,296,000
Wavelength bandpass (μm)	0.757-0.772, 1.59-1.62, 2.04-2.08	0.758-0.775, 1.56-1.72, 1.92-2.08, 5.56-14.3	0.24-0.44, 0.4-1.0, 1.0-1.7, 1.94-2.04, 2.265-2.38	3.74-4.61, 6.20-8.22, 8.80-15.4	3.62-5.0, 5.0-8.26, 8.26-15.5
Signal/noise (nadir, 5% albedo)	>300 @ 1.59-1.62 μm, >240 @ 2.04-208 μm	~120 @ 1.56-1.72 μm, ~120 @ 1.92-2.08	<100 @ 1.57 μm	~2,000 @ 4.2 μm, ~1,400 @ 3.7-13.6 μm, ~800 @ 13.6-15.4 μm	~1,000 @ 12 μm, ~500 @ 4.5 μm
Orbit altitude	705 km	666 km	790 km	705 km	820 km
Local time	13:30 ± 0:1.5	13:00 ± 0:15	10:00	13:30	21:30
Revisit time/orbits	16 days/233 orbits	3 days/72 orbits	35 days	16 days/233 orbits	72 days/1,037 orbits
Launch date	Failed on launch	January 2009	March 2002	May 2002	October 2006
Nominal life	2 years	5 years	7+ years	7+ years	5 years

NOTES: AIRS = Atmospheric Infrared Sounder; CAI = Cloud and Aerosol Imager; FTS = Fourier transform spectrometer; GOSAT = Greenhouse gases Observing Satellite; IASI = Infrared Atmospheric Sounding Interferometer; OCO = Orbiting Carbon Observatory; SCIAMACHY = Scanning Imaging Absorption Spectrometer for Atmospheric Chartography; SWIR = short-wavelength infrared; TIR = thermal infrared.

[a]Crisp (2008); Crisp et al. (2008).
[b]Akihiko Kuze, Japan Aerospace Exploration Agency, personal communication, 2009; Hamazaki et al. (2007); Shiomi et al. (2007).
[c]<http://envisat.esa.int/instruments/sciamachy/>; Burrows et al. (1995); Noël et al. (1998); Buchwitz et al. (2005).
[d]Aumann et al. (2003); Chahine et al. (2008).
[e]Phulpin et al. (2007); Crevoisier et al. (2009).
[f]Instantaneous field-of-view/Swath.
[g]The uncertainty represents the estimate of random errors (e.g., the effects of detector noise) and additional systematic errors (e.g., bias caused by cloud and aerosol effects) unaccounted for or otherwise eliminated from the total error. Bias is reduced by successful validation efforts.

The GOSAT uncertainty is dominated by the precision (random errors). For OCO, Crisp et al. (2004) and Miller et al. (2007) discuss the observational system simulation experiments, including modeling of the OCO instrument performance characteristics, that led to an instrument design that would meet a measurement requirement of 1 part per million (ppm). The as-built OCO instrument performance was verified during prelaunch tests, which included direct solar observations. The analysis of the latter gave the best confirmation that the as-built instrument performance exceeded its design requirements.

The methods for bias reduction and validation are the same for GOSAT and OCO. Washenfelder et al. (2006) demonstrated the OCO validation concept and the essential role of ground-based measurements for meeting those objectives. Bösch et al. (2006) used these ground-based measurements to validate SCIAMACHY CO_2. The GOSAT team also plans to use the same validation sites and instruments. OCO planned to include and use Aeronet measurements. The OCO validation plan purposely located ground-based validation measurements at Atmospheric Radiation Measurement (ARM) Program sites to capitalize on the wealth of ancillary atmospheric and surface measurements.

can be directly measured on shorter time scales. Time trends in oceanic CO_2 at a single point are illustrated in Figure C.1. Most well-qualified oceanic CO_2 datasets reside at the Department of Energy's Carbon Dioxide Information Analysis Center.[1]

Early work to track the accumulated burden of anthropogenic CO_2 in seawater relied on tracer data, such as bomb ^{14}C. The use of tracers was necessary because of the high natural background level of dissolved CO_2 in seawater, the complexity of the processes affecting its distribution, and the relatively small size of the anthropogenic signal, all of which combined to make direct observation an uncertain business. Today, there are widely available accurate standards and mea-

---
[1]See <http://cdiac.ornl.gov/>.

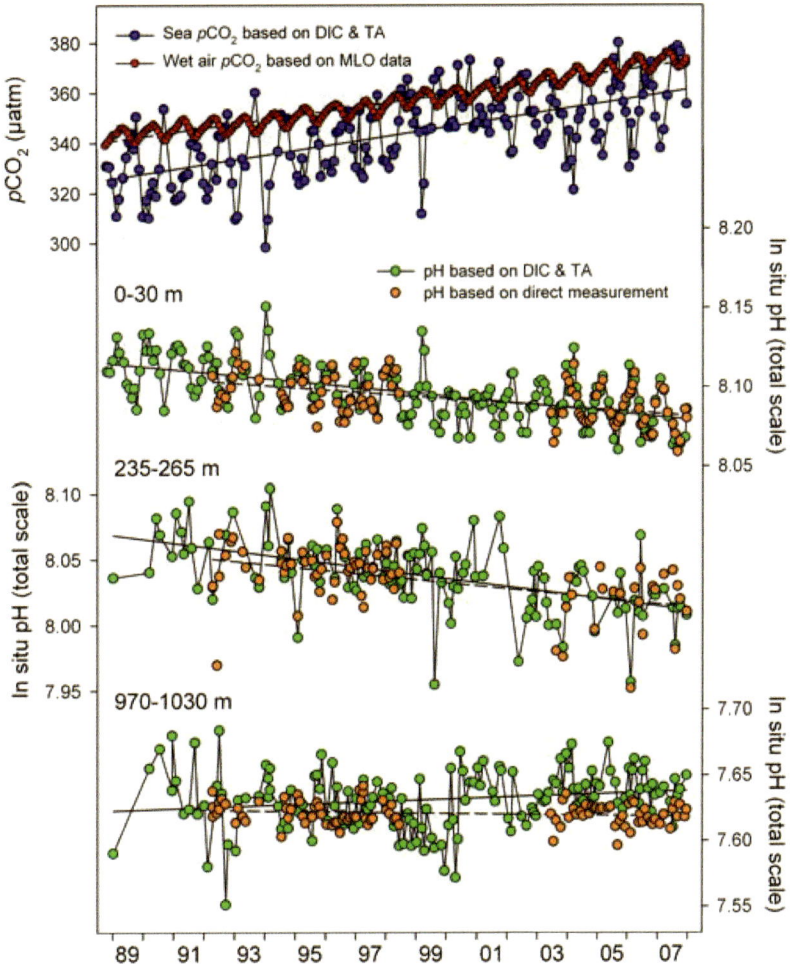

FIGURE C.1 Time trend of surface water pCO$_2$ offshore Hawaii, showing the direct tracking with atmospheric CO$_2$ forcing and the resultant change in ocean pH. The change in pH results from reaction with dissolved carbonate ion and causes a decline in the buffer capacity of seawater. The penetration to depth can also be seen in the changing subsurface data. SOURCE: Dore et al. (2009). Copyright 2009 National Academy of Sciences, U.S.A.

surements, greatly improved knowledge of the functioning of natural cycles, and an enormous increase in the anthropogenic CO$_2$ signal.

The first demonstrated recovery of the anthropogenic CO$_2$ signal from direct ocean measurements was by Brewer (1978), who corrected for the subsurface changes in dissolved CO$_2$ due to respiration and carbonate dissolution and showed that the residual pCO$_2$ signal closely resembles the atmospheric CO$_2$ history of the water mass. An additional term to correct for local air-sea disequilibrium at the water mass source was applied by Gruber et al. (1996), and techniques such as these are widely used today. In addition, comparison of datasets from different cruise years now allows simple tracking of the changing ocean anthropogenic CO$_2$ burden. An example of the ability to record the increasing storage of anthropogenic CO$_2$ in the ocean is shown in Figure C.2.

Methane

The chemistry of ocean methane (CH$_4$) is complex (see the review by Reeburgh, 2007); determining the extent to which the atmosphere is affected and detecting and understanding regional changes (e.g., ocean basin scale, or preferably less) are considerable challenges. First, the global methane budget contains significant oceanic terms (Table C.2). The net ocean emissions to the atmosphere are only about 2 percent of the total, mostly because large amounts of methane originating in

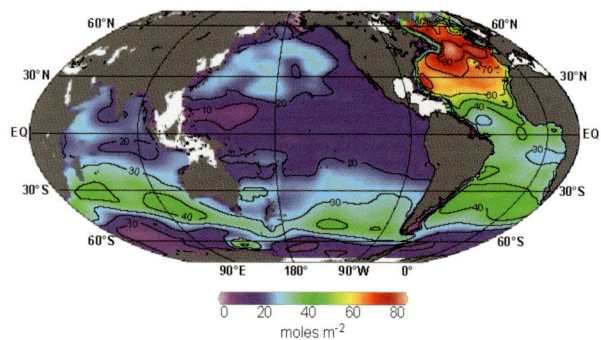

FIGURE C.2 Column inventory of anthropogenic CO_2 in the ocean as of 1994. The accumulated burden is 388 ± 62 billion tons CO_2 and is growing at a rate of ~7.4 billion tons per year. Thus, the inventory in 2009 is ~500 billion tons CO_2. SOURCE: Figure 1 from Sabine et al. (2004). Reprinted with permission from AAAS.

continental margin sediments are consumed by microbial processes before they can be released into the fluid ocean. The terms for methane hydrate decomposition and release and for tracing the signature of gas plumes emitted from the seafloor, which can affect regional signals, are being updated rapidly.

Ocean Water Column. The first ocean water column measurements of methane were made in the late 1960s (Swinnerton and Linnenbom, 1967), revealing nanomolar concentrations and values well below atmospheric equilibrium at depth, indicating oceanic consumption within the water column. Measurements of oceanic profiles by Scranton and Brewer (1977) showed the puzzling existence of a significant maximum in concentration just below the oceanic mixed layer, a pattern later found over large regions of the global ocean (Watanabe et al., 1995). The puzzle was likely solved by Karl et al. (2008), who documented aerobic production of methane by decomposition of methyl phosphonate when consumed by phytoplankton in phosphate-starved environments. This process likely accounts for the small net source of CH_4 to the atmosphere from the upper ocean (Table C.2). Thus, observations of a peak in methane concentrations in the upper ocean should not be confused with industrial releases.

TABLE C.2 Global Net Methane Emissions

Source or Sink	Emissions (Tg CH_4 yr^{-1})	Consumption (Tg CH_4 yr^{-1})	Gross Production (Tg CH_4 yr^{-1})
Animals	80	0	80
Wetlands	115	27	142
Bogs, tundra (boreal)	35	15	50
Swamps, alluvial	80	12	92
Rice production	100	477	577
Biomass burning	55	0	55
Termites	20	24	44
Landfills	40	22	62
Oceans, freshwaters	10	75.3	85.3
Hydrates	5?	5	10
Coal production	35	0	35
Gas production	40	18	58
Venting, flaring	10	0	10
Distribution leaks	30	18	48
Total sources	500		
Chemical destruction	−450		
Soil consumption	−10	40	40
Total sinks	−460	688.3	
Total production			1,188.3

NOTE: Tg = terragrams – million metric tons
SOURCE: Reeburgh (2003).

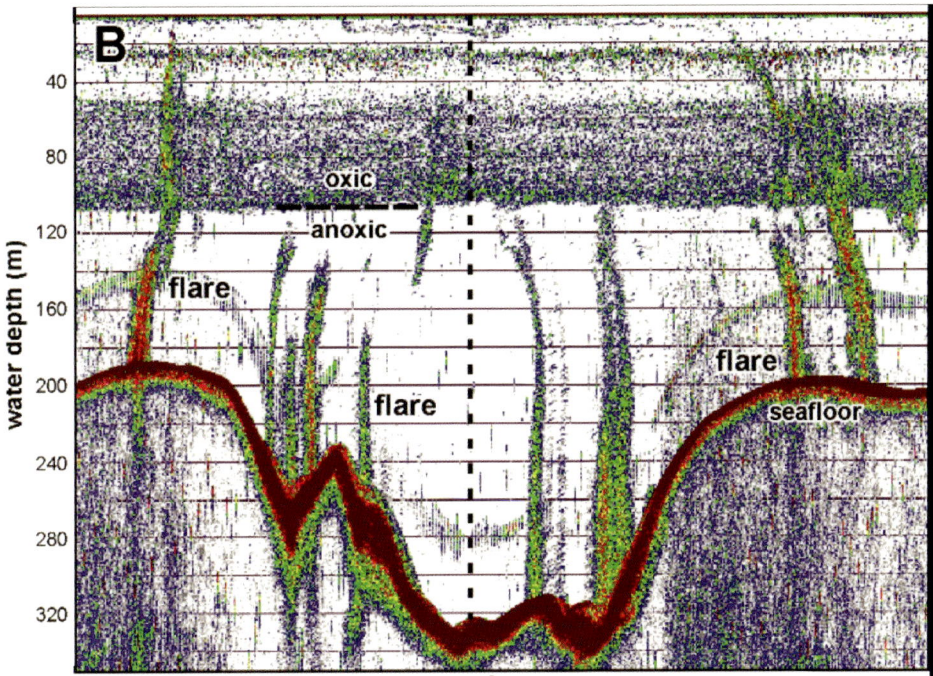

FIGURE C.3 Acoustic signatures of methane plumes rising from the floor of the Black Sea offshore from the Crimean peninsula. SOURCE: McGinnis et al. (2006). Copyright 2006 American Geophysical Union. Reproduced by permission of the American Geophysical Union.

Methane in the upper ocean, whether transferred there by exchange with the atmosphere or created locally, is consumed through oxidation to CO_2 as the water masses are transferred to depth (Scranton and Brewer, 1978). The recent rise in atmospheric CH_4 concentrations is imprinted on this process (Rehder et al., 1999); methane originating in the deep sea around vents is also quickly consumed.

Venting of Methane from the Seafloor. Methane is vented naturally from the large reservoirs on the continental shelves, but little reaches the atmosphere. For example, Figure C.3 shows the acoustic detection of a field of methane plumes rising from the floor of the Black Sea (Schmale et al., 2005; McGinnis et al., 2006). Although the height of the plumes (some are higher than 1,300 m) is impressive, little of the gas is vented to the atmosphere. The reason is that gas bubbles venting from the seafloor become coated with a film of hydrate ($CH_4.6H_2O$) or oily material from the higher hydrocarbons, which slows the dissolution rate of the rising bubble by about a factor of 4 (Rehder et al., 2002). The Black Sea is an unusual case because the deep water is anoxic and the density contrast between deep and surface layers is very strong. Nonetheless, both observations and models show that even the largest plumes undergo such significant dissolution during their rise to the surface that only plumes originating from very shallow sources (~100 m depth) can provide a source of oceanic CH_4 to the atmosphere.

Satellites. The use of satellites to detect and quantify methane releases from the ocean has received little attention. The most novel and useful approach was taken by MacDonald and colleagues (MacDonald et al., 2008), who used synthetic aperture radar (SAR) imagery of the Gulf of Mexico to provide a basin-wide inventory of gas seeps based upon their surface expression. The team identified some 1,821 sources and estimated the methane emissions to the atmosphere. The effect seen in the SAR imagery (Figure C.4) is the damping of capillary waves from the trace oil residue carried on the rising bubbles and reaching the sea surface. A pure methane gas stream with zero associated

C.2). Although vast quantities of hydrates are known to occur in nature, recent estimates range from 500-2,500 Gt C (Milkov, 2004) to 63,400 Gt C (Klauda and Sandler, 2005). A map of known hydrate locations is shown in Figure C.5.

The potential for global warming to destabilize seafloor methane hydrates has been debated since the early 1980s (e.g., Revelle, 1983). The quantities are so large that destabilization of hydrates could be of grave concern. Fortunately, the danger seems small; almost all of the methane released from seafloor hydrates would simply dissolve into the surrounding water and then be microbially oxidized to CO_2 (Hester and Brewer, 2009).

Nitrous Oxide

The oceans are a source of N_2O to the atmosphere, emitting some 25-33 percent of the total flux (Hirsch et al., 2006). However, using this information to assign fluxes to any specific region, or decoding the oceanic component of trends over reasonable periods of time (a decade or so), will be exceptionally difficult. The concentration of dissolved N_2O in seawater is nanomolar.

The source of N_2O in the ocean is intimately

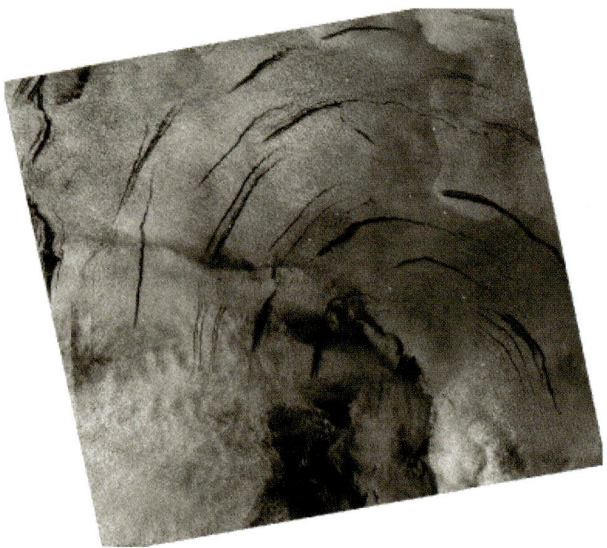

FIGURE C.4 SAR image of the Gulf of Mexico showing the surface expression of gas seeps from the trace oil components. SOURCE: MacDonald et al. (2008). Copyright 2008 American Geophysical Union. Reproduced by permission of the American Geophysical Union.

higher hydrocarbons would not show this effect, but the oil-gas association is very common.

Methane Hydrates. Releases of CH_4 from methane hydrates are uncertain (see the question mark in Table

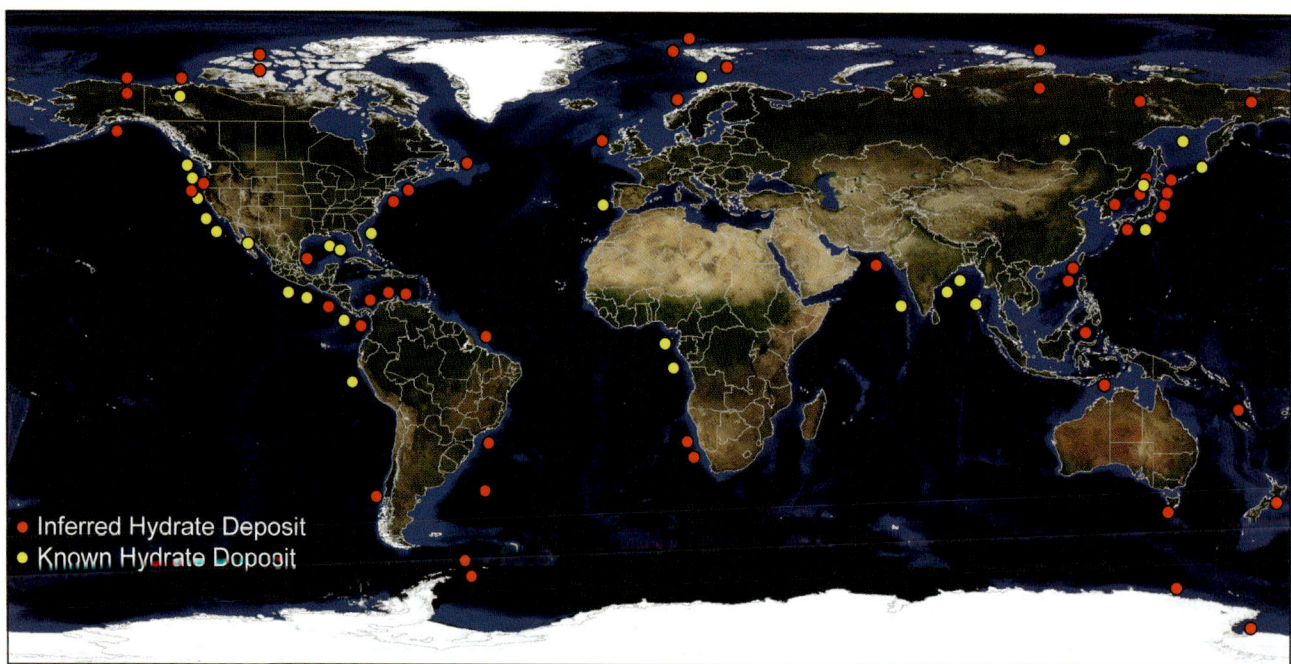

FIGURE C.5 Worldwide map of more than 90 hydrate occurrences; such sites could be monitored from space for evidence of methane releases. SOURCE: Hester and Brewer (2009). Reproduced with permission of Annual Reviews, Inc.; permission conveyed through Copyright Clearance Center, Inc.

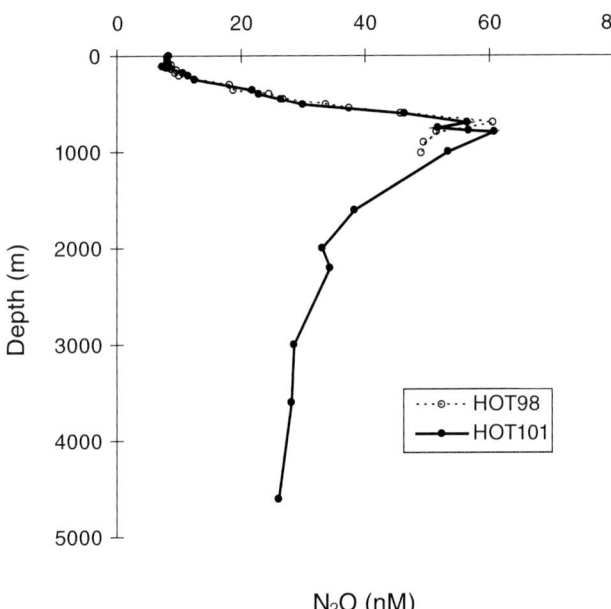

FIGURE C.6 Typical oceanic depth profile of dissolved N_2O, observed at the Hawaii Ocean Time Series station. The presence of higher levels of N_2O is strongly correlated with the oxygen minimum. SOURCE: Reprinted from Ostrom et al. (2000) with permission from Elsevier.

involved with microbial transformations of nitrogen compounds and shows strong spatial heterogeneity. The intense subsurface low-oxygen regions of the northern Indian Ocean have been shown to be a significant source (Law and Owens, 1990) of N_2O production, but transfer of the gas to the atmosphere occurs erratically with wind-driven events. Stable isotope evidence suggests that the well-oxygenated upper water column in subtropical waters is a large source of atmospheric N_2O (Dore et al., 1998), but the broad pattern is consistent with production of N_2O in low-oxygen regions during the nitrification process (Nevison et al., 2003). A plot of a typical oceanic profile is shown in Figure C.6.

It is widely recognized that under a warming climate scenario, deep ocean oxygen levels will decline. Thus, models have used the relationship between N_2O and O_2 to predict strongly increased N_2O fluxes from the future ocean (Schmittner et al., 2008; Shaffer et al., 2009). For example, Schmittner et al. (2008) estimate that by the year 4000, atmospheric N_2O will increase by about 21 percent as a result of this increased oceanic flux as a weak positive feedback. However, these very long term effects are likely to have little impact on twenty-first century treaty monitoring.

REFERENCES

Aumann, H.H., M.T. Chahine, C. Gautier, M.D. Goldberg, E. Kalnay, L.M. McMillin, H. Revercomb, P.W. Rosenkranz, W.L. Smith, D.H. Staelin, L.L. Strow, and J. Susskind, 2003, AIRS/AMSU/HSB on the Aqua Mission: Design, science objectives, data products, and processing systems, *IEEE Transactions on Geoscience and Remote Sensing*, **41**, 253.

Bösch, H., G.C. Toon, B. Sen, R.A. Washenfelder, P.O. Wennberg, M. Buchwitz, R. deBeek, J.P. Burrows, D. Crisp, M. Christi, B.J. Connor, V. Natraj, and Y.L. Yung, 2006, Space-based near-infrared CO_2 measurements: Testing the OCO retrieval algorithm and validation concept using SCIAMACHY observations over Park Falls, Wisconsin, *Journal of Geophysical Research*, **111**, D23302, doi:10.1029/2006JD007080.

Brewer, P.G., 1978, Direct observation of the oceanic CO_2 increase, *Geophysical Research Letters*, **5**, 997-1000.

Buchwitz, M., R. de Beek, S. Noël, J.P. Burrows, H. Bovensmann, H. Bremer, P. Bergamaschi, S. Körner, and M. Heimann, 2005, Carbon monoxide, methane and carbon dioxide columns retrieved from SCIAMACHY by WFM-DOAS: Year 2003 initial data set, *Atmospheric Chemistry and Physics*, **5**, 3313-3329.

Burrows, J.P., E. Hölzle, A.P.H. Goede, H. Visser, and W. Fricke, 1995, SCIAMACHY—Scanning Imaging Absorption Spectrometer for Atmospheric Chartography, *Acta Astronautica*, **35**, 445-451.

Chahine, M., C. Barnet, E.T. Olsen, L. Chen, and E. Maddy, 2005, On the determination of atmospheric minor gases by the method of vanishing partial derivatives with application to CO_2, *Geophysical Research Letters*, **32**, L22803, doi:10.1029/2005GL024165.

Chahine, M.T., L. Chen, P. Dimotakis, X. Jiang, Q. Li, E.T. Olsen, T. Pagano, J. Randerson, and Y.L. Yung, 2008, Satellite remote sounding of mid-tropospheric CO_2, *Geophysical Research Letters*, **35**, L17807, doi:10.1029/2008GL035022.

Conway, T.J., P.P. Tans, L.S. Waterman, K.W. Thoning, D.R. Kitzis, K.A. Masarie, and N. Zhang, 1994, Evidence for interannual variability of the carbon cycle from the NOAA/GMCC global air sampling network, *Journal of Geophysical Research*, **99**, 22,831-22,855.

Crevoisier, C., S. Heilliette, A. Chedin, S. Serrar, R. Armante, and N.A. Scott, 2004, Midtropospheric CO_2 concentration retrieval from AIRS observations in the tropics, *Geophysical Research Letters*, **31**, L17106, doi:10.1029/2004GL020141.

Crevoisier, C., A. Chedin, H. Matsueda, T. Machida, R. Armante, and N.A. Scott, 2009, First year of upper tropospheric integrated content of CO_2 from IASI hyperspectral infrared observations, Discussion, *Atmospheric Chemistry and Physics*, **9**, 8187-8222.

Crisp, D., 2008, The Orbiting Carbon Observatory: NASA's first dedicated carbon dioxide mission, in *Sensors, Systems, and Next-Generation Satellites XII*, Proceedings of SPIE, **7106**, 710604.

Crisp, D., R.M. Atlas, F.-M. Breon, L.R. Brown, J.P. Burrows, P. Ciais, B.J. Connor, S.C. Doney, I.Y. Fung, D.J. Jacob, C.E. Miller, D. O'Brien, S. Pawson, J.T. Randerson, P. Rayner, R.J. Salawitch, S.P. Sander, B. Sen, G.L. Stephens, P.P. Tans, G.C. Toon, P.O. Wennberg, S.C. Wofsy, Y.L. Yung, Z. Kuang, B. Chudasama, G. Sprague, B. Weiss, R. Pollock, D. Kenyon, and S. Schroll, 2004, The Orbiting Carbon Observatory (OCO) mission, *Advances in Space Research*, **34**, 700-709.

Crisp, D., C.E. Miller, and P.L. DeCola, 2008, NASA Orbiting Carbon Observatory: Measuring the column averaged carbon dioxide mole fraction from space, *Journal of Applied Remote Sensing*, **2**, 023508, doi:10.1117/1.2898457.

Dore, J.E., B.N. Popp, D.M. Karl, and F.J. Sansone, 1998, A large source of atmospheric nitrous oxide from subtropical North Pacific surface waters, *Nature*, **396**, 63-65.

Dore, J.E., R. Lukas, D.W. Sadler, M.J. Church, and D.M. Karl, 2009, Physical and biogeochemical modulation of ocean acidification in the central North Pacific, *Proceedings of the National Academy of Sciences*, **106**, 12,235-12,240.

Gruber, N., J.L. Sarmiento, and T.F. Stocker, 1996, An improved method for detecting anthropogenic CO_2 in the oceans, *Global Biogeochemical Cycles*, **10**, 809-837.

Hamazaki, T., Y. Kaneko, A. Kuze, and H. Suto, 2007, Greenhouse gases observation from space with TANSO-FTS on GOSAT, in *Fourier Transform Spectroscopy/Hyperspectral Imaging and Sounding of the Environment*, Optical Society of America Technical Digest Series, paper FWB1.

Hester, K.C., and P.G. Brewer, 2009, Clathrate hydrates in nature, *Annual Review of Marine Science*, **1**, 303-327.

Hirsch, A.I., A.M. Michalak, L.M. Bruhwiler, W. Peters, E.J. Dlugokencky, and P.P. Tans, 2006, Inverse modeling estimates of the global nitrous oxide surface flux from 1998-2001, *Global Biogeochemical Cycles*, **20**, GB1008, doi:10.1029/2004GB002443.

Karl, D.M., L. Beversdorf, K.M. Bjorkman, M.J. Church, A. Martinez, and E.F. DeLong, 2008, Aerobic production of methane in the sea, *Nature Geoscience*, **1**, 473-478.

Klauda, J.B., and S.I. Sandler, 2005, Global distribution of methane hydrate in ocean sediment, *Energy & Fuels*, **19**, 459-470.

Law, C.S., and N.J.P. Owens, 1990, Significant flux of atmospheric nitrous oxide from the northwest Indian Ocean, *Nature*, **346**, 826-828.

MacDonald, I.R., V. Asper, O. Garcia, M. Kastner, I. Leifer, T.H. Naehr, E. Solomon, S. Yvon-Lewis, and B. Zimmer, 2008, HyFlux—Part 1: Regional modeling of methane flux from near-seafloor gas hydrate deposits on continental margins, American Geophysical Union, fall meeting, abstract OS33A-1322.

Maddy, E.S., C.D. Barnet, M. Goldberg, C. Sweeney, and X. Liu, 2008, CO_2 retrievals from the Atmospheric Infrared Sounder: Methodology and validation, *Journal of Geophysical Research*, **113**, D11301, doi:10.1029/2007JD009402.

McGinnis, D.F., J. Greinert, Y. Artemov, S.E. Beaubien, and A. Wüest, 2006, Fate of rising methane bubbles in stratified waters: How much methane reaches the atmosphere? *Journal of Geophysical Research*, **111**, C09007, doi:10.1029/2005JC003183.

Milkov, A.V., 2004, Global estimates of hydrate-bound gas in marine sediments: How much is really out there? *Earth-Science Reviews*, **66**, 193-197.

Miller, C.E., D. Crisp, P.L. DeCola, S.C. Olsen, J.T. Randerson, A.M. Michalak, A. Alkhaled, P. Rayner, D.J. Jacob, P. Suntharalingam, D.B.A. Jones, A.S. Denning, M.E. Nicholls, S.C. Doney, S. Pawson, H. Bösch, B.J. Connor, I.Y. Fung, D. O'Brien, R.J. Salawitch, S.P. Sander, B. Sen, P. Tans, G.C. Toon, P.O. Wennberg, S.C. Wofsy, Y.L. Yung, and R.M. Law, 2007, Precision requirements for space-based XCO_2 data, *Journal of Geophysical Research*, **112**, D10314, doi:10.1029/2006JD007659.

Nevison, C., J.H. Butler, and J.W. Elkins, 2003, Global distribution of N_2O and the ΔN_2O-AOU yield in the subsurface ocean, *Global Biogeochemical Cycles*, **17**, 1119, doi:10.1029/2003GB002068.

Noël, S., H. Bovensmann, J.P. Burrows, J. Frerick, K.V. Chance, A.P.H. Goede, and C. Muller, 1998, The SCIAMACHY instrument on ENVISAT-1, in *Sensors, Systems, and Next-Generation Satellites II*, Proceedings of SPIE, **3498**, 94-104.

Ostrom, N.E., M.E. Russ, B. Popp, T.M. Rust, and D.M. Karl, 2000, Mechanisms of nitrous oxide production in the subtropical North Pacific based on determinations of the isotopic abundances of nitrous oxide and di-oxygen, *Chemosphere—Global Change Science*, **2**, 281-290.

Phulpin, T., D. Blumstein, F. Prel, B. Tournier, P. Prunet, and P. Schlüssel, 2007, Applications of IASI on MetOp-A: First results and illustration of potential use for meteorology, climate monitoring, and atmospheric chemistry, in *Atmospheric and Environmental Remote Sensing Data Processing and Utilization III: Readiness for GEOSS*, Proceedings of SPIE, **6684**, 66840F.

Prinn, R.G., P.G. Simmonds, R.A. Rasmussen, R.D. Rosen, F.A. Alyea, C.A. Cardelino, A.J. Crawford, D.M. Cunnold, P.J. Fraser, and J.E. Lovelock, 1983, The Atmospheric Lifetime Experiment 1, Introduction, instrumentation, and overview, *Journal of Geophysical Research*, **88**, 8353-8367.

Reeburgh, W.S., 2003, Global methane biogeochemistry, in *The Atmosphere, Treatise on Geochemistry*, **4**, R.F. Keeling, H.D. Holland, and K.K. Turekian, eds., Elsevier-Pergamon, Oxford, U.K., pp. 65-89.

Reeburgh, W.S., 2007, Oceanic methane biogeochemistry, *Chemical Reviews*, **107**, 486-513.

Rehder, G., R. Keir, E. Suess, and M. Rhein, 1999, Methane in the Northern Atlantic controlled by microbial oxidation and atmospheric history, *Geophysical Research Letters*, **26**, 587-590.

Rehder, G., P.G. Brewer, E.T. Peltzer, and G. Friederich, 2002, Enhanced lifetime of methane bubble streams within the deep ocean, *Geophysical Research Letters*, **29**, 10.1029/2001GL013966.

Revelle, R.R., 1983, Methane hydrates in continental slope sediments and increasing carbon dioxide, in *Changing Climate*, National Academy Press, Washington, D.C., pp. 252-261.

Sabine, C.L., R.A. Feely, N. Gruber, R.M. Key, K. Lee, J.L. Bullister, R. Wanninkhof, C.S. Wong, D.W.R. Wallace, B. Tilbrook, F.J. Millero, T.-H. Peng, A. Kozyr, T. Ono, and A.F. Rios, 2004, The oceanic sink for anthropogenic CO_2, *Science*, **305**, 367-371.

Schmale, O., J. Greinert, and G. Rehder, 2005, Methane emission from high-intensity marine gas seeps into the atmosphere, *Geophysical Research Letters*, **32**, L07609, doi:10.1029/2004GL021138.

Schmittner, A., A. Oschlies, H.D. Matthews, and E.D. Galbraith, 2008, Future changes in climate, ocean circulation, ecosystems, and biogeochemical cycling simulated for a business-as-usual CO_2 emissions scenario until year 4000 AD, *Global Biogeochemical Cycles*, **22**, GB1013, doi:10.1029/2007GB002953.

Schneising, O., M. Buchwitz, J.P. Burrows, H. Bovensmann, M. Reuter, J. Notholt, R. Macatangay, and T. Warneke, 2008, Three years of greenhouse gas column-averaged dry air mole fractions retrieved from satellite—Part 1: Carbon dioxide, *Atmospheric Chemistry and Physics*, **8**, 3827-3853.

Scranton, M.I., and P.G. Brewer, 1977, Occurrence of methane in the near-surface waters of the western subtropical North Atlantic, *Deep-Sea Research*, **24**, 127-138.

Scranton, M.I., and P.G. Brewer, 1978, Consumption of dissolved methane in the deep ocean, *Limnology and Oceanography*, **23**, 1207-1213.

Shaffer, G., S.M. Olsen, and J.O.P. Pedersen, 2009, Long-term ocean oxygen depletion in response to carbon dioxide emissions from fossil fuels, *Nature Geoscience*, **2**, 105-109.

Shiomi, K., S. Kawakami, T. Kina, Y. Mitomi, M. Yoshida, N. Sekio, F. Kataoka, and R. Higuchi, 2007, Calibration of the GOSAT sensors, in *Sensors, Systems, and Next-Generation Satellites XI*, Proceedings of SPIE, **6744**, 67440G.

Strow, L.L., and S.E. Hannon, 2008, A 4-year zonal climatology of lower tropospheric CO_2 derived from ocean-only atmospheric infrared sounder observations, *Journal of Geophysical Research*, **113**, D18302, doi:10.1029/2007JD009713.

Swinnerton, J.W., and V.J. Linnenbom, 1967, Determination of C1 to C4 hydrocarbons in seawater by gas chromatography, *Journal of Gas Chromatography*, **5**, 570-573.

Washenfelder, R.A., G.C. Toon, J.-F. Blavier, Z. Yang, N.T. Allen, P.O. Wennberg, S.A. Vay, D.M. Matross, and B.C. Daube, 2006, Carbon dioxide column abundances at the Wisconsin Tall Tower site, *Journal of Geophysical Research*, **111**, D22305, doi:10.1029/2006JD007154.

Watanabe, S., N. Higashitani, N. Tsurushima, and S. Tsunogai, 1995, Methane in the western North Pacific, *Journal of Oceanography*, **51**, 39-60.

Appendix D

Technologies for Measuring Emissions by Large Local Sources

New technologies that have great potential for measuring the local increase in greenhouse gases around large point sources include the same multispecies gas analyzers that are suitable for commercial aircraft applications and spectrometers similar to the Total Carbon Column Observing Network (TCCON). Concurrent measurements of wind speed, wind direction, humidity, solar radiation, and boundary-layer height would allow adjustments for seasonal and interannual variability in atmospheric mixing. When "simple," low-cost trace gas sensors or remote sensing tools are used, one has to be careful that air density and water vapor variations are not misinterpreted as changes of the trace gas dry air mole fraction, which is the only property containing information about recent sources. Mobile as well as fixed trace gas measurement approaches are likely to be important in characterizing urban to remote gradients and may be particularly useful for identifying point sources of methane (CH_4), nitrous oxide (N_2O), and halocompounds. For non-carbon dioxide (CO_2) greenhouse gases, an urban network may provide new first-order information about different source terms—again allowing for more effective policy that targets specific industries or practices.

Airborne lidar (light detection and ranging) techniques offer the potential for monitoring greenhouse gas emissions downwind of sources. An aircraft-deployed differential absorption lidar (DIAL), flown within or just above the boundary layer across an emission plume downwind of the source, can measure the dimensions and column content of gases within and outside of the plume. Combining the in-plume concentration measurements with estimates of the wind, obtained via models or measured directly from the aircraft, enables a calculation of the emission rate of gases exported from the source. The technique has been demonstrated for estimation of ozone fluxes in plumes downwind of urban areas and oil refineries (e.g., Senff et al., 2006) using observations from a downward-looking lidar flown on a low-flying National Oceanic and Atmospheric Administration (NOAA) Twin Otter and wind extrapolated from wind profiler measurements.

DIAL techniques have been shown to be potentially feasible for measuring both profiles and column content of important greenhouse gases. Ehret et al. (2008) estimated errors in measurements of CO_2, CH_4, and N_2O columns from satellite-borne DIAL instruments for global monitoring. They examined DIAL systems operating in several absorbing infrared spectral regions and concluded that such measurements were indeed feasible, with systematic errors estimated to be less than 0.4 percent for CO_2, 0.6 percent for CH_4, and 0.3 percent for N_2O. Because DIAL observations from low-flying aircraft are significantly easier than satellite measurements due to much shorter distance to surface and the lack of interference from high- and mid-level clouds, the aircraft measurement problem is potentially feasible with much smaller instruments than those assumed for the satellite study. Note that lidar measures concentration, not mole fraction. The measurements rely on the dry air density being the same inside and outside of a plume and on very well characterized surface elevations. These measurements are thus local and cannot readily be compared with mole fraction measurements at other places and times.

Research efforts are under way, primarily within the National Aeronautics and Space Administration (NASA), to demonstrate airborne DIAL measurements of CO_2 as a step toward eventual space-based measurements. Browell et al. (2008) and Abshire et al. (2009) have both reported on airborne measurements of CO_2 optical thickness at 1.57 μm. To avoid the problem of having to convert a column absorption measurement of a trace gas to a mole fraction, one can deploy a second lidar that simultaneously measures the column amount of oxygen (O_2), which is a very accurate indicator of the total amount of dry air in the column. The challenges in developing high-precision DIAL systems for column content measurements are associated primarily with instrumentation issues such as maintaining and monitoring long-term laser stability. However, given the resources currently being directed toward the problem in both the United States and Europe, there is a high likelihood that such challenges will be met.

Hardesty et al. (2008) described co-deployment of a Doppler wind lidar and water vapor DIAL instrument on a research aircraft to measure horizontal transport of moisture over the southern Great Plains. In this way, airborne lidar techniques can be used as improvements over application of models for observing wind characteristics within the plume. The accuracy of the lidar wind measurement is about 10 cm s^{-1}. These combined techniques could thus be applied to obtain measurements of greenhouse gas emission without the need for application of models.

REFERENCES

Abshire, J.B., H. Riris, G.R. Allan, C. Weaver, J. Mao, and W. Hasselbrack, 2009, Airborne lidar measurements of atmospheric CO_2 column absorption and line shapes from 3-11 km altitudes, *Geophysical Research Abstracts*, **11**, EGU2009-11507.

Browell, E.V., M.F. Dobbs, J. Dobler, S. Kooi, Y. Choi, F.W. Harrison, B. Moore III, and T.S. Zaccheo, 2008, Airborne demonstration of 1.57-micron laser absorption spectrometer for atmospheric CO_2 measurements, in *Proceedings of the 24th International Laser Radar Conference*, Boulder, Colo., January 12-15, pp. 697.

Ehret, G., C. Kiemle, M. Wirth, A. Amediek, A. Fix, and S. Houweling, 2008, Space-borne remote sensing of CO_2, CH_4, and N_2O by integrated path differential absorption lidar: A sensitivity analysis, *Applied Physics B*, **90**, 593-608.

Hardesty, R.M., W.A. Brewer, C.J. Senff, B.J. McCarty, G. Ehret, A. Fix, C. Kiemle, and E.I. Tollerud, 2008, Structure of meridional moisture transport over the U.S. southern Great Plains observed by co-deployed airborne wind and water vapor lidars, in *Symposium on Recent Developments in Atmospheric Applications of Radar and Lidar*, American Meteorological Society, January 21-24.

Senff, C.J., R.J. Alvarez II, R.M. Hardesty, R.M. Banta, L.S. Darby, A.M. Weickmann, S.P. Sandberg, D.C. Law, R.D. Marchbanks, W.A. Brewer, D.A. Merritt, and J.L. Machol, 2008, Airborne lidar measurements of ozone flux and production downwind of Houston and Dallas, in *Proceedings of the 24th International Laser Radar Conference*, Boulder, Colo., June 23-27, pp. 659-662.

Appendix E

Biographical Sketches of Committee Members

Stephen W. Pacala, *chair*, is Frederick D. Petrie Professor of Ecology and Evolutionary Biology at Princeton University and director of the Princeton Environmental Institute. He also co-directs the Carbon Mitigation Initiative, a collaboration between Princeton University, British Petroleum, and the Ford Motor Company to develop strategies to reduce global carbon dioxide emissions. Dr. Pacala received his Ph.D. in biology from Stanford University. His research focuses on ecology and modeling, with an emphasis on the interactions between greenhouse gases, climate, and the biosphere. He was a coordinating lead author of a chapter on the North American carbon budget in the 2006 assessment *The First State of the Carbon Cycle Report: The North American Carbon Budget and Implications for the Global Carbon Cycle* (CCSP, 2007). Among his many honors are the David Starr Jordan Prize and the George Mercer Award of the Ecological Society of America. Dr. Pacala is a fellow of the American Association for the Advancement of Science and a member of the American Academy of Arts and Sciences and the National Academy of Sciences.

Clare Breidenich is an independent consultant with more than 12 years of experience on climate change policy in general and on the Kyoto Protocol of the United Nations Framework Convention on Climate Change (UNFCCC) in particular. From 2002 to 2006, she was a senior program officer at the UNFCCC Secretariat, where she managed the review process for national greenhouse gas inventories of 40 countries and directed activities related to data systems and procedures for the Kyoto Protocol's reporting, review and compliance procedures. This experience, as well as work for the Environmental Protection Agency (EPA) and the State Department has given her extensive knowledge of the technical and policy options for greenhouse gas mitigation, including market mechanisms, and methodologies and protocols for estimation, reporting, and verification of greenhouse gas emissions and reductions. Ms. Breidenich has an M.S. in environmental science from Indiana University and a B.A. from the University of Michigan.

Peter G. Brewer is an ocean chemist and senior scientist at the Monterey Bay Aquarium Research Institute. His research interests include the ocean chemistry of greenhouse gases, the geochemistry of gas hydrates, ocean acidification, and the evolution of the oceanic fossil-fuel CO_2 signal. He has devised novel techniques for measuring and extracting the oceanic signatures of global change. Dr. Brewer has served on many committees associated with ocean trace gases, including the Joint Global Ocean Fluxes Committee, the National Research Council (NRC) Panel on Policy Implications of Greenhouse Gas Warming: Mitigation, and the Scientific Committee on Oceanic Research's Working Group 75 on Ocean CO_2 Monitoring. He was a member of MEDEA. He is a fellow of the American Association for the Advancement of Science and of the American Geophysical Union, serving as president of the Ocean Sciences Section for 2 years. He received a Ph.D. and a B.S. from Liverpool University in England.

Inez Fung is a professor of atmospheric sciences and founding co-director of the Berkeley Institute of the Environment at the University of California, Berkeley. She studies the interactions between climate change and biogeochemical cycles, particularly the processes that maintain and alter the composition of the atmosphere. Her research emphasis is on using atmospheric transport models and a coupled carbon-climate model to examine how CO_2 sources and sinks are changing. She was also a member of the science team for the National Aeronautics and Space Administration's (NASA's) Orbiting Carbon Observatory (OCO). Dr. Fung is a recipient of the American Geophysical Union's Roger Revelle Medal and appears in a new National Academy of Sciences biography series for middle-school readers *Women's Adventure in Science*. She is a fellow of the American Meteorological Society and the American Geophysical Union and a member of the National Academy of Sciences. She received her B.S. in applied mathematics and her Ph.D. in meteorology from the Massachusetts Institute of Technology (MIT).

Michael R. Gunson is an atmospheric scientist and the chief scientist of the Jet Propulsion Laboratory's Earth Science and Technology Directorate. His research interests focus on understanding the physical and chemical processes of the Earth's atmosphere using space-based instruments. He is the deputy principal investigator of NASA's Tropospheric Emission Spectrometer, which measures the radiance emitted by Earth's surface and by gases and particles in Earth's atmosphere. The data are used to study air quality and transport of pollution around the globe. Dr. Gunson was awarded several NASA exceptional service medals for his leadership and scientific achievements associated with space-based instruments that measure atmospheric radiative transfer, chemistry, and physical processes. He received a Ph.D. and a B.S. in chemistry from Bristol University.

Gemma Heddle is the carbon management adviser with Corporate Health, Environment, and Safety at Chevron. She is responsible for managing the development and deployment of Chevron's new energy and emissions inventory system, for revising Chevron's greenhouse gas emissions reporting protocol, for leading the European Union and U.S. focus areas of the company's carbon markets team, and for managing Chevron's internal carbon trading registry. Prior to joining Chevron in 2005, she worked as a technology policy analyst at the Cooperative Research Centre for Greenhouse Gas Technologies (CO2CRC) in Australia. Ms. Heddle holds a dual M.S. in technology and policy and in civil and environmental engineering from MIT, an M.S. in chemical engineering from the University of Sydney, Australia, and a double B.S. and B.A. from the University of Adelaide, Australia.

Beverly E. Law is a professor of global change forest science in the College of Forestry at Oregon State University. Her research focuses on the role of forests, woodlands, and shrublands in the global carbon cycle. Her approach is interdisciplinary, involving observations and models to study changes in climate, management, and other land-use changes that influence carbon and water cycling across a region over seasons to decades. Dr. Law is the science chair of the AmeriFlux network, which provides continuous observations of ecosystem level exchanges of CO_2, water, and energy at more than 100 research sites in the Americas. She is a member of the Science Steering Group of the U.S. Carbon Cycle Science Program and the Science Steering Committee of the North American Carbon Program. She also serves as the U.S. point of contact on scientific exchanges in carbon cycle science for State Department bilateral agreements with Italy, Canada, and the European Union. She received a Ph.D. in forest science from Oregon State University and a B.S. in forest resources and conservation from the University of Florida.

Gregg Marland is a senior research staff member in the Environmental Sciences Division at Oak Ridge National Laboratory. In addition to research on the sources of greenhouse gas emissions and mitigation options, he helped define the methodologies and emissions coefficients now used to estimate CO_2 emissions to the atmosphere. Dr. Marland served on the NRC Panel on Policy Implications of Greenhouse Warming and has been a lead author on several reports of the Intergovernmental Panel on Climate Change (IPCC). He has spent recent sabbatical years as guest professor at Mid Sweden University in Östersund and as senior research scholar at the International Institute for Applied Systems Analysis in Laxenburg, Austria. Dr.

Marland received a Ph.D. in geology from the University of Minnesota and a B.S. from Virginia Tech.

Keith Paustian is a professor of soil ecology in the Department of Soil and Crop Sciences and a senior research scientist in the Natural Resources Ecology Laboratory at Colorado State University. His main fields of interest include agroecosystem ecology, soil organic matter dynamics, and global change. He is currently leading projects to assess soil carbon sequestration in several states and to develop national inventories of carbon emissions and sequestration. His research also involves the development of ecosystem and economic assessments to advise policy makers on climate change mitigation. He is a leader on the IPCC and the Council for Agricultural Science and Technology task force on agricultural mitigation of greenhouse gases. He is an editor of a recent book *Soil Organic Matter in Temperate Agroecosystems: Long-Term Experiments in North America* (CRC Press, 1997).

Michael J. Prather is the Fred Kavli Chair and Professor in the Department of Earth System Science at the University of California, Irvine. From 2005 to 2006, he was a Jefferson science fellow at the State Department. His research focuses on simulation of the physical, chemical, and biological processes that determine atmospheric composition, including global chemical transport models that describe ozone and other trace gases. Dr. Prather has played a leading role in international assessments of ozone and climate change. He has served on numerous NRC committees and chaired the Planning Group for the Workshop on Direct and Indirect Human Contributions to Terrestrial Greenhouse Gas Fluxes. He is a fellow of the American Geophysical Union and the American Association for the Advancement of Science and is a member of the Norwegian Academy of Science and Letters. Dr. Prather received a Ph.D. in astronomy and astrophysics from Yale and undergraduate degrees in mathematics from Yale and physics from Merton, Oxford.

James T. Randerson is a professor in the Department of Earth System Science at the University of California, Irvine. Dr. Randerson uses trace gas observations from ground- and space-based instruments and models to study the global carbon cycle. He is currently investigating pathways of rapid carbon loss from terrestrial ecosystems, including fire emissions and permafrost degradation. He is a member of the science team for NASA's Orbiting Carbon Observatory and co-chair of the biogeochemistry working group for the Community Climate System Model. Dr. Randerson is a fellow of the American Geophysical Union and a recipient of the James B. Macelwane Medal. He received a Ph.D. in biological sciences and a B.S. in chemistry from Stanford University.

Pieter P. Tans is senior scientist at the Earth System Research Laboratory of the National Oceanic and Atmospheric Administration (NOAA) in Boulder, Colorado. His research interests focus on inverse models and data assimilation, atmospheric chemistry and transport, carbon cycle, and global climate change. His group maintains the world's largest global monitoring network of atmospheric greenhouse gas concentrations and provides reference gas mixtures to calibrate high-accuracy greenhouse gas measurements worldwide. Dr. Tans has served on several advisory committees related to the carbon cycle and climate. He has received several medals from the Department of Commerce and is a fellow of the American Geophysical Union. He received a Ph.D. in physics and a doctorandus (roughly equivalent to an M.S.) in theoretical physics from Rijksuniversiteit Groningen, The Netherlands.

Steven C. Wofsy is the Abbott Lawrence Rotch Professor of Atmospheric and Environmental Sciences at Harvard University. His work focuses on the chemical composition of the atmosphere, using data analysis and modeling to understand sources, sinks, transformations, and transport of atmospheric trace gases. His research group also develops airborne sensors to make accurate measurements of CO_2, CH_4, CO, and N_2O. He has chaired or been a member of several carbon cycle and NRC advisory committees. Dr. Wofsy is a recipient of the American Geophysical Union's James B. Macelwane Award and NASA's Distinguished Public Service Medal. He is a fellow of the American Geophysical Union and the American Association for the Advancement of Science. He received a Ph.D. in chemistry from Harvard and a B.S. in chemistry from the University of Chicago.

Appendix F

Acronyms and Abbreviations

AFOLU	agriculture, forestry, and other land use		Range Weather Forecasts Integrated Forecast System
AIRS	Atmospheric Infrared Sounder	EDGAR	Emission Database for Global Atmospheric Research
ALE-GAGE	Atmospheric Lifetime Experiment-Global Atmospheric Gases Experiment	ESRL	Earth System Research Laboratory
ASCENDS	Active Sensing of CO_2 Emissions over Nights, Days, and Seasons	FAO	UN Food and Agriculture Organization
ASTER	Advanced Spaceborne Thermal Emission and Reflection Radiometer	FIA	U.S. Forest Service Inventory and Analysis
AVHRR	Advanced Very High Resolution Radiometer	GAW	Global Atmosphere Watch
		GDP	gross domestic product
AVIRIS	Airborne Visible/Infrared Imaging Spectrometer	GEMS	Global and Regional Earth-System Monitoring Using Satellite and In situ Data
CFC	chlorofluorocarbon		
CH_4	methane	GOFC-GOLD	Global Observations of Forest and Land Cover Dynamics
CO_2	carbon dioxide		
COP	Conference of Parties (United Nations)	GOSAT	Greenhouse gases Observing Satellite
CTMs	chemistry-transport models	Gt	billion metric tons
		GTOS-TCO	Global Terrestrial Observing System-Terrestrial Carbon Observations
DESDynI	Deformation, Ecosystem Structure and Dynamics of Ice		
DETER	Real-Time Detection of Deforestation	HFC	hydrofluorocarbon
DIAL	differential absorption lidar	HiFIS	High-fidelity Imaging Spectrometer
DOC	degradable organic carbon		
DOE	Department of Energy		
		IASI	Infrared Atmospheric Sounding Interferometer
ECMWF IFS	European Centre for Medium-		

ICESat-II	Ice, Cloud, and Land Elevation Satellite-II	PFC	perfluorocarbon
ICOS	Integrated Carbon Observation System	ppb	parts per billion
		ppm	parts per million
IEA	International Energy Agency	ppt	parts per trillion
INTEX/NA	Intercontinental Chemical Transport Experiment—North America	REDD	Reducing Emissions from Deforestation and Forest Degradation in Developing Countries
IPCC	Intergovernmental Panel on Climate Change		
IPPU	industrial processes and product use	SAR	synthetic aperture radar
		SCIAMACHY	Scanning Imaging Absorption Spectrometer for Atmospheric Chartography
kWh	kilowatt-hour		
		SF_6	sulfur hexafluoride
LDCM	Landsat Data Continuity Mission	SMAP	Soil Moisture Active-Passive
lidar	light detection and ranging	SSM/I	Special Sensor Microwave Imager
LUCF	land-use change and forestry	SWDS	solid waste disposal site
MODIS	Moderate Resolution Imaging Spectrometer	TCCON	Total Carbon Column Observing Network
MOPITT	Measurements of Pollution in the Troposphere	TES	Tropospheric Emission Spectrometer
Mt	million tons	Tg	million metric tons
		TRACE-P	Transport and Chemical Evolution over the Pacific
NASA	National Aeronautics and Space Administration		
NOAA	National Oceanic and Atmospheric Administration	UNFCCC	United Nations Framework Convention on Climate Change
NO_x	nitrogen oxide	USDA	U.S. Department of Agriculture
N_2O	nitrous oxide		
NWP	numerical weather prediction	WMO	World Meteorological Organization
O_2	oxygen		
OCO	Orbiting Carbon Observatory	WRF	Weather Research and Forecast model
ORCA	Oregon and California		
OSSE	Observation System Simulation Experiment		